## PRAISE FOR *FLU*

'Fascinating scientific history.' *Guardian*

'Kolata dramatically conveys the shivers virologists feel when they get their hands on 1918 genes.' *The Times*

'A page-turner . . . well researched and well plotted.' *Financial Times*

'Compelling and informative . . . Kolata writes with the authority of an expert . . . in a lucid and accessible style.' *Irish News* Book of the Week

'Big ideas, high stakes and the aftermath of tragedy on a colossal scale.' *Times Educational Supplement*

'She commands the intelligent curiosity, well-honed reporting techniques and smooth prose style of a top science reporter.' *Glasgow Sunday Herald*

'Read this book to recalibrate your perspective.' *British Medical Journal*

'Her readable and compelling narrative is highly recommended.' *New York Times*

## PRAISE FOR GINA KOLATA'S *CLONE*

'[An] intelligent, wide-ranging account of cloning's science, history and ethics.' *Observer*

'This gripping account of science in action keeps you wanting to know what happens next.' *Independent*

'A thoughtful, engaging, interpretative account . . . there is much in the book that most experts can learn from.' *New Scientist*

'Superb but unsettling account of the cloning of Dolly the sheep.' J. G. Ballard, *Sunday Times*

# FLU

Gina Kolata has been writing for the *New York Times* for twelve years and before that was a senior writer for *Science* magazine. She is the author of *Clone: The Road to Dolly and the Path Ahead* and *The Baby Doctors: Probing the Limits of Fetal Medicine*, and is co-author of the landmark *Sex in America*. She has won numerous awards for her writing on science, medicine and mathematics.

## ALSO BY GINA KOLATA

*Clone: The Road to Dolly and the Path Ahead*

*The Baby Doctors: Probing the Limits of Fetal Medicine*

*Sex in America: A Definitive Survey* (with Robert T. Michael, John H. Gagnon and Edward O. Laumann)

# FLU

## THE STORY OF THE GREAT INFLUENZA PANDEMIC OF 1918 AND THE SEARCH FOR THE VIRUS THAT CAUSED IT

GINA KOLATA

PAN BOOKS

First published 1999 by Farrar, Straus and Giroux, New York
First published in Great Britain 2000 by Macmillan

This edition published 2001 by Pan Books
an imprint of Macmillan Publishers Ltd
25 Eccleston Place, London SW1W 9NF
Basingstoke and Oxford
Associated companies throughout the world
www.macmillan.com

ISBN 0 330 48423 0



1 3 5 7 9 8 6 4 2

A CIP catalogue record for this book is available from
the British Library.

Printed and bound in Great Britain by
Mackays of Chatham plc, Chatham, Kent

*For my parents*

# CONTENTS

# PROLOGUE

If anyone should have known about the 1918 flu, it was I.

I was a microbiology major in college and even took a course in virology. But the 1918 flu was never mentioned. I also took history courses in college, with one of my favorites being a class that covered the important events of the twentieth century. But although World War I was a major part of the course, the 1918 flu was not discussed. I have written about diseases and medicine for my entire career, first at *Science* magazine and then at *The New York Times*, even writing articles about influenza. But I never paid any attention to the 1918 flu.

In retrospect, it is hard for me to understand my ignorance. The 1918 flu epidemic puts every other epidemic of this century to shame. It was a plague so deadly that if a similar virus were to strike today, it would kill more people in a single year than heart disease, cancers, strokes, chronic pulmonary disease, AIDS, and Alzheimer's disease combined. The epidemic affected the course of history and was a terrifying presence at the end of World War I, killing more Americans in a single year than died in battle in World War I, World War II, the Korean War, and the Vietnam War.

The 1918 flu even affected my family and my husband's family. My father insisted on heeding the advice of an old doctor who had lived through that flu and who decided, as a consequence, to blast every respiratory illness with erythromycin. As a child, I took the antibiotic every time I had a fever, useless though it was in fighting most common respiratory diseases. Yet I did not make the connection with the doctor's frightening experience with the 1918 flu and his steadfast faith in a wonder drug discovered decades later. When I grew older and understood the overuse of antibiotics, I would disparage my father's doctor, arguing that he was irrational.

In my husband's family, the flu had been a life-altering event. My husband's mother was a young girl when her father died of the viral infection and her mother was left to bring up four children alone. Somehow, though, neither my husband nor I quite realized what had happened. His mother had always said that her father died of pneumonia that he caught from working in a foundry.

It seems remarkable to me now that I never grasped the fact that a terrible epidemic had swept the world in 1918, leaving death and devastation in its wake and touching nearly every family with its icy hand. But, I have learned, I was not alone in my ignorance. The 1918 influenza epidemic is one of history's great conundrums, obliterated from the consciousness of historians, who traditionally ignore science and technology but not, for the most part, plagues.

My epiphany came in 1997, when I wrote an article in *The New York Times* about a remarkable paper being published in the journal *Science*. That paper, which involved the first attempts to resurrect the virus's genetic code, was also a key clue in a medical mystery story that is as astonishing as the 1918 influenza itself. It involves science and politics, at their most confused and at their finest. It involves a virus that is one of the worst killers ever known. And it involves researchers who became obsessed with tracking the virus down. Like all good mystery stories, it also has elements of serendipity and surprise.

It is a story that begged to be told, both for the sheer drama of the tale and for its implications. The resolution of the mystery could help scientists save humanity if that terrible virus or another one like it stalks the earth again.

# FLU

# 1

# THE PLAGUE YEAR

*This is a detective story. Here was a mass murderer that was around 80 years ago and who's never been brought to justice. And what we're trying to do is find the murderer.*

—Jeffery Taubenberger, molecular pathologist

When the plague came, on those chilly days of autumn, some said it was a terrible new weapon of war. The plague germs were inserted into aspirin made by the German drug company Bayer. Take an aspirin for a headache and the germs will creep through your body. Then your fate is sealed.

No, the plague came in on a camouflaged German ship that had crept into Boston Harbor under cover of darkness and released the germs that seeded the city. Boston, after all, was where the plague started. There was an eyewitness, an old woman who said she saw a greasy-looking cloud that floated over the harbor and wafted over the docks.

No, it was started by Germans who slipped into Boston Harbor on U-boats and then sneaked ashore, carrying vials of the plague germs with them. They let the germs loose in theaters and among crowds gathered for those interminable Liberty Bond rallies. Lieutenant Colonel Philip S. Doane, head of the Health Sanitation Section of the Emergency Fleet Corporation, said so, and he certainly was in a position to know. It was on page one of the *Philadelphia Inquirer*.

Soon the plague was everywhere. And no one was safe.

The sickness preyed on the young and healthy. One day you are fine, strong, and invulnerable. You might be busy at work in your office. Or maybe you are knitting a scarf for the brave troops fighting the war to end all wars. Or maybe you are a soldier reporting for basic training, your first time away from home and family.

You might notice a dull headache. Your eyes might start to burn. You start to shiver and you will take to your bed, curling up in a ball. But no amount of blankets can keep you warm. You fall into a restless sleep, dreaming the distorted nightmares of delirium as your fever climbs. And when you drift out of sleep, into a sort of semi-consciousness, your muscles will ache and your head will throb and you will somehow know that, step by step, as your body feebly cries out "no," you are moving steadily toward death.

It may take a few days, it may take a few hours, but there is nothing that can stop the disease's progress. Doctors and nurses have learned to spot the signs. Your face turns a dark brownish purple. You start to cough up blood. Your feet turn black. Finally, as the end nears, you frantically gasp for breath. A blood-tinged saliva bubbles out of your mouth. You die—by drowning, actually—as your lungs fill with a reddish fluid.

And when a doctor does an autopsy, he will observe your lungs lying heavy and sodden in your chest, engorged with a thin bloody liquid, useless, like slabs of liver.

They called the plague of 1918 influenza, but it was like no influenza ever seen before. It was more like a biblical prophecy come true, something from Revelations that predicted that first the world was to be struck by war, then famine, and then, with the breaking of the fourth seal of the scroll foretelling the future, the appearance of a horse, "deathly pale, and its rider was called Plague, and Hades followed at its heels."

The plague took off in September of that year, and when it was over, half a million Americans would lie dead. The illness spread to the most remote parts of the globe. Some Eskimo villages were decimated, nearly eliminated from the face of the earth. Twenty percent of Western Samoans perished. And no matter where it struck, the virus went after an unusual group—young adults who generally are spared the ravages of infectious diseases. The death curves were W-shaped, with peaks for the babies and toddlers under age 5, the elderly who were aged 70 to 74, and people aged 20 to 40.

Children were orphaned, families destroyed. Some who lived through it said it was so horrible that they would not even talk about it. Others tried to put it behind them as another wartime nightmare, somehow conflating it with the horrors of trench warfare and mustard gas. It came when the world was weary of war. It swept the globe in months, ending when the war did. It went away as mysteriously as it appeared. And when it was over, humanity had been struck by a disease that killed more people in a few months' time than any other illness in the history of the world.

When we think of plagues we think of strange and terrible illnesses. AIDS. Ebola. Anthrax spores. And, of course, the Black Death. We worry about horrifying symptoms—pustules, or fountains of blood gushing from every orifice. Or young men, who had had the bodies of gods, reduced to skeletal figures, hobbling down the street on withered limbs, leaning on canes, shivering with cold. Today we worry about germ warfare—a new virus made of a combination of smallpox and anthrax or smallpox and Ebola. Or we worry that a terrifying new disease is brewing somewhere, in a hot zone, and that it is poised, prepared, with the disruption of ancient forests, to break out and kill us all.

But influenza never makes the list of deadly plagues. It seems so innocuous. It comes around every winter and everyone gets it sooner or later. There is no good treatment once a person becomes ill, but no matter. Nearly everyone gets over it, few are the worse for the experience. It is just an inconvenient illness that inflicts, at most, a week or so of misery. Influenza is not supposed to be deadly, at least for young adults, who have little reason to fear death or disease.

Even its name, "influenza," hints at its usual pattern of coming around each winter. "Influenza" is an Italian word that, one hypothesis has it, was coined by the disease's Italian victims in the middle of the eighteenth century. *Influenza di freddo* means "influence of the cold."

Flu also, however, seems unavoidable. It is spread through the air and there is little that can be done to prevent being infected. "I know how not to get AIDS," says Alfred W. Crosby, a historian of the 1918 flu. "I don't know how not to get the flu."

And yet perhaps because the flu is so familiar, its terrors in 1918 were all the more dreadful. It is like a macabre science fiction tale in which the mundane becomes the monstrous.

When the illness was first observed, doctors were reluctant even to call it the flu. It seemed to be a new disease, they said. Some called it bronchopneumonia, others called it epidemic respiratory infection. Doctors suggested it might be cholera or typhus, or perhaps it was dengue fever or botulism. Still others said it was simply an unidentified pandemic disease. Those who used the term "influenza" insisted on enclosing it in quotation marks.

One way to tell the story of the 1918 flu is through facts and figures, a collection of data whose impact is numbing and whose magnitude is almost inconceivable.

How many became ill? More than 25 percent of the U.S. population.

What about servicemen, the very young and healthy who were the virus's favorite targets? The Navy said that 40 percent of its members got the flu in 1918. The Army estimated that about 36 percent of its members were stricken.

How many died worldwide? Estimates range from 20 million to more than 100 million, but the true number can never be known. Many places that were bludgeoned by the flu did not keep mortality statistics, and even in countries such as the United States, efforts at tabulating flu deaths were complicated by the fact that there was no definitive test in those days to show that a person actually had the flu. But still, the low end of the mortality estimates is stunning. In comparison, AIDS had killed 11.7 million people through 1997. World War I was responsible for 9.2 million combat deaths and around 15 million total deaths. World War II for 15.9 combat deaths. Historian Crosby remarks that whatever the exact number felled by the 1918 flu, one thing is indisputable: the virus "killed more humans than any other disease in a period of similar duration in the history of the world."

How lethal was it? It was twenty-five times more deadly than ordinary influenzas. This flu killed 2.5 percent of its victims. Normally, just one-tenth of 1 percent of people who get the flu die. And since a fifth of the world's population got the flu that year, including 28 percent of Americans, the number of deaths was stunning. So many died, in fact, that the average life span in the United States fell by twelve years in 1918. If such a plague came today, killing a similar fraction of the U.S. population, 1.5 million Americans would die, which is more than the number felled in a single year by heart disease, cancers, strokes, chronic pulmonary disease, AIDS, and Alzheimer's disease combined.

But the raw numbers cannot convey the scenes of horror and misery that swept the world in 1918, which became part of everyday life in every nation, in the largest cities and remotest hamlets.

Some tell of their personal epiphanies. Historian Crosby, a friendly bear of a man with snow-white hair and a short bushy beard, was at Washington State University one day, gazing at a wall of world almanacs. On a whim, he picked up an almanac from 1917 and looked up the U.S. life expectancy. It was, he recalls, about fifty-one years. He then turned to the almanac from 1919. The life expectancy was about the same. Then he looked at the 1918 almanac. The life expectancy was thirty-nine years, he says. "What the hell happened? The life expectancy had dropped to what it had been fifty years before." Then he realized what the explanation must be. It was the influenza epidemic, the flu his own father had lived through but had not spoken about to Crosby. "When you talk to people who lived through it, they think it was just their block or just their neighborhood," Crosby noted. The flu's enormous, almost unthinkable impact somehow had escaped attention. Crosby applied for a grant from the National Institutes of Health to study the 1918 flu and soon became the world's expert on that almost forgotten period of history.

No one knows for sure where the 1918 flu came from or how it turned into such a killer strain. All that is known is that it began as an ordinary flu but then it changed. It infected people in the spring of 1918, sickening its victims for about three days with chills and fever, but rarely killing them. Then it disappeared, returning in the fall with the power of a juggernaut.

In retrospect, medical experts talk of the two waves of the 1918 flu. The first was banal, and easily forgotten. No one mentioned plagues or germ warfare when the influenza epidemic first arrived. But when it came back, in the second wave, it had become something monstrous, bearing little resemblance to what is ordinarily thought of as the flu.

The earliest traces of the first wave of the 1918 flu are lost in the sands of time, a warning that only afterward seemed dire. The disease seemed trivial at the time, coming as it did in the midst of the disruptions and terrors of war. But for one of the first towns to be hit by the flu, the disease was not easily dismissed—not because it was so deadly but because it was so infectious.

It was February and the tourist season was in full swing in San Sebastián. The sunny town on the northern coast of Spain seemed a world apart from the dismal, dreary fighting just over the border in France. San Sebastián in the winter of 1918 was a place where you might forget the trenches and the wet, cold, muddy battles. You could escape the talk of mustard gas, that deadly green haze, that horrible new German weapon of war. You could find respite in a country that was still unaligned, where the days were warm and the nights soft and fragrant. You might forget that the rest of Europe was bogged down in an exhausting war to end all wars.

Then the flu came to town. It was nothing alarming—just three days or so of fever, aches, and pains. But it certainly was contagious. It seemed that nearly everyone who was exposed to the disease became ill about two days later. And the disease seemed to strike young, healthy adults, often sparing the old people and the children, who usually are the first to be felled by influenza.

What to do? If the world knew about the flu in San Sebastián, the tourist season would be finished. Who would want to go on vacation only to be laid up with the flu? Maybe the illness could be hushed up, the town's officials reasoned. Yet the word spread—San Sebastián was a place to be avoided.

At nearly the same time, some soldiers were becoming ill, although there was as yet no clear pattern of the disease's spread. Influenza arrived in March in the 15th U.S. Cavalry traveling to Europe.

Two months later, it seemed that everyone was getting sick. In Spain, eight million people were ill, including King Alfonso XIII. One-third of Madrid was sick with the flu, forcing some government offices to close. Even the trams stopped running. And this time, Spain was not alone—the first wave of the flu had spread widely.

The troops called it "three-day fever," according to some who suffered from it. One, Sergeant John C. Acker of the 107th Ammunition Train, 32nd Division, American Expeditionary Force, writing from France that April, said: "They started calling it the 'three-day fever' here, but couldn't camouflage it with a name when it runs its course in a week or more. It hits suddenly and one's temperature nearly chases the mercury thru the top of the M.D.'s thermometer, face gets red, every bone in the body aches and the head splits wide open. This continues for three or four days and then disappears after considerable perspiration, but the 'hangover' clings for a week or two."

But in the rest of the world, the illness came to be called the Spanish flu, to Spain's consternation. After all, the other countries of Europe, as well as the United States and countries in Asia, were hit too in that spring of 1918. Maybe the name stuck because Spain, still unaligned, did not censor its news reports, unlike other European countries. And so Spain's flu was no secret, unlike the flu elsewhere.

Nonetheless, the scope of the epidemic remains unclear. There were no requirements in those days to report cases of influenza—that became a practice in the United States only after, and as a consequence of, the second wave of the 1918 flu. And there was no reason in those days of war to keep track of what seemed like a minor illness. Reports on the flu's reaches were sporadic, reflecting mostly the practices of organizations such as prisons, the military, and some industries, which simply recorded absentees. There was no systematic attempt to track an epidemic.

There are records noting that at the Ford Motor Company more than 1,000 workers called in sick with the flu in March. In San Quentin prison, 500 of 1,900 prisoners became ill in April and May. On March 4, the flu came to Camp Funston (now Fort Riley) in Kansas, a training camp for 20,000 recruits. That month and the next, it also arrived at more than a dozen other Army camps, but no eyebrows were raised. After all, colds and flu were to be expected in training camps where thousands of men were brought together, mingling and passing viruses among themselves.

In April 1918, the epidemic appeared in France, laying waste to British, American, and French troops stationed there, as well as the civilian population. The next month, it was in England, where King George V got the flu. The epidemic crested in England in June; at the same time, it cropped up in China and Japan. In Asia, it also was called the "three-day fever" or, sometimes, "wrestler's fever."

Not surprisingly, the epidemic affected the war effort. Soldiers trying to fight in World War I were laid up by the flu in such numbers that some commanders complained that the disease was hindering their ability to fight.

King George's Grand Fleet could not even put to sea for three weeks in May, with 10,313 men sick. The British Army's 29th Division had planned to attack La Becque on June 30, but had to put off the operation because so many of its men were sick with the flu.

German General Erich von Ludendorff, the leader of the country's acclaimed offense, complained that the flu, or the Flanders fever, as the Germans called it, was thwarting his battle plans. It was not enough that the fighting men were hungry and cold and wet, trying to slog their way through fields of mud that could swallow a tank. Now there was this flu which, Ludendorff said, was weakening his men and lowering their morale. The flu, he added, contributed to the failure of his July offensive, a battle plan that nearly won the war for Germany.

He also groused about his staff's complaints about the flu. "It was a grievous business having to listen every morning to the Chiefs of Staff's recital of the number of influenza cases, and their complaints about the weakness of their troops."

Yet although much of the world fell ill that spring, there remained large areas that were untouched. Most of Africa and almost all of South America and Canada had no flu epidemic. And as summer arrived, even the countries that were hardest hit had a reprieve. The flu seemed to vanish without a trace.

But a few months later the flu was back with a vengeance.

It roared into the world, beginning in places where its impact was, at first, not widely known. The second wave of the 1918 pandemic still was highly contagious. But this time it was a killer. Its path was evident in retrospect, when demographers traced patterns of unusually high death rates among young adults. By August, notes Gerald Pyle, a medical geographer at the University of North Carolina, the disease "had grimly cut its swath among populations of the Indian subcontinent, Southeast Asia, Japan, China, a large part of the Caribbean, and parts of Central and South America."

Although about 20 percent of its victims had a mild disease and recovered without incident, the rest had one of two terrifying illnesses. Some almost immediately became deathly ill, unable to get enough oxygen because their lungs had filled with fluid. They died in days, or even hours, delirious with a high fever, gasping for breath, lapsing at last into unconsciousness. In others, the illness began as an ordinary flu, with chills, fever, and muscle aches, but no untoward symptoms. By the fourth or fifth day of the illness, however, bacteria would swarm into their injured lungs and they would develop pneumonia that would either kill them or lead to a long period of convalescence.

The second wave of the flu arrived in the United States in Boston, appearing among a group of sailors who docked at the Commonwealth Pier in August. The sailors were simply in transit, part of the vast movement of troops in a war that transformed daily life.

By then, the war effort had taken over the country. No man wanted to be left behind—the worst thing you could call a man was a slacker. And so a quarter of Americans had signed up to fight, with those men who remained behind embarrassed, apologizing for medical conditions that kept them from the front. The women spent their days visiting hospitals, bringing bright baskets of flowers and sweets, and wrapping bandages for the men abroad.

And then some of those sailors in Boston got sick.

On August 28, eight men got the flu. The next day, 58 were sick. By day four, the sick toll reached 81. A week later, it was 119, and that same day the first civilian was admitted to Boston City Hospital sick with the flu.

Deaths soon followed. On September 8, three people died from the flu in Boston: a Navy man, a merchant marine, and a civilian.

That same day, the flu appeared in Fort Devens, Massachusetts, thirty miles west of Boston.

Overnight, Fort Devens became a scene out of hell. One doctor, assigned to work in the camp that September, wrote despairingly to a friend about an epidemic that was out of control. The doctor's letter is dated September 29, 1918, signed with his first name, "Roy." Nothing more is known about who he was or what became of him. His letter was discovered more than sixty years later, in a trunk in Detroit, and was published in the *British Medical Journal* in December 1979, having been sent in by a Scottish doctor, N. R. Grist of the University of Glasgow, who saw it as a cautionary tale.

Roy wrote: "Camp Devens is near Boston, and has about 50,000 men, or did before this epidemic broke loose." The flu epidemic hit the camp four weeks earlier, he added, "and has developed so rapidly that the camp is demoralized and all ordinary work is held up till it has passed. All assemblages of soldiers are taboo."

The disease starts out looking like an ordinary sort of influenza, Roy explained. But when the soldiers are brought to the hospital at the Army base, they "rapidly develop the most viscous type of Pneumonia that has ever been seen. Two hours after admission they have the Mahogany spots over the cheek bones and a few hours later you can begin to see the Cyanosis extending from the ears and spreading all over the face, until it is hard to distinguish the colored man from the white. It is only a matter of a few hours then until death comes and it is simply a struggle for air until they suffocate. It is horrible. One can stand to see one, two, or twenty men die, but to see these poor devils dropping like flies gets on your nerves. We have been averaging about 100 deaths a day, and still keeping it up."

It became a problem just to dispose of the dead. "It takes Special trains to carry away the dead," Roy remarked. "For several days there were not coffins and the bodies piled up something fierce and we used to go down to the morgue (which is just back of my ward) and look at the boys laid out in long rows. It beats any sight they ever had in France after a battle. An extra long barracks has been vacated for the use of the Morgue, and it would make any man sit up and take notice to walk down the long lines of soldiers all dressed and laid out in double rows. We have no relief here, you get up in the morning at 5:30 and work steady until about 9:30 P.M., sleep, then go at it again."

Even medical experts were shocked by what they saw at Camp Devens. Just six days before Roy wrote his letter, on September 23, the U.S. Surgeon General had sent one of the nation's leading doctors to the camp to figure out what was going on. The doctor, William Henry Welch, was a pathologist, a scientist, and a physician whose honors were almost unprecedented. He had been president of the most prestigious science and medical societies, including the American Medical Association, the National Academy of Science, and the American Association for the Advancement of Science. Some said that he was as highly regarded in his time as Benjamin Franklin had been earlier.

But Welch was as unprepared as Roy for the 1918 flu. He had, in fact, thought that American troops were astonishingly healthy. In September 1918, Welch, along with Colonel Victor C. Vaughan, who also had been president of the American Medical Association, Dr. Rufus Cole, president of the Rockefeller Institute, and Simeon Walbach of Harvard Medical School, had just completed a tour of inspections of Army camps in the South. They were flush with the success of the public health effort that seemed to have made disease in the military almost a thing of the past. Welch, in fact, had concluded that the camps were in such good condition, the troops in such good health, that he could retire. He was seventy-one years old, a portly, genial bachelor, and felt he had completed his service to his country. Then he was called upon to investigate the carnage that was occurring at Fort Devens.

The four doctors were summoned to Washington to speak to the U.S. Surgeon General, Dr. William C. Gorgas, the man who had eliminated yellow fever from Cuba. Gorgas called the men into his office, barely glancing up from the papers on his desk when the distinguished team came in. Then he said, "You will proceed immediately to Devens. The Spanish influenza has struck that camp."

The doctors obeyed, of course, traveling to Union Station near the Capitol building in Washington and boarding the next train to Fort Devens. They arrived the following morning on a dreary day, when a frigid rain was falling and dying soldiers, sodden and chilled, were filing into the hospital, carrying their blankets, burning with fever, shivering with cold, and coughing up bloody mucus.

What the doctors saw horrified them. The camp, built for 35,000 men, was overcrowded with 45,000. And the influenza epidemic was running rampant. In just twenty-four hours preceding Welch's visit, 66 men had died. The day that Welch and his retinue came, 63 died. The hospital, built to hold 2,000, was overflowing with 8,000 men.

Vaughan wrote about the experience. He was a man who had seen epidemics before. He had seen typhoid fever and seen firsthand how that illness felled men in the Spanish-American War. But never had he seen anything like the influenza epidemic in Fort Devens, Massachusetts.

Vaughan said there was no point in discussing the history of the influenza epidemic, which "visited the remotest corners, taking toll of the most robust, sparing neither soldier nor civilian, and flaunting its red flag in the face of science." But the scene at Devens was one he could never forget.

When the group of medical officials arrived, they saw what Vaughan said was inscribed in his brain. These memories are "ghastly ones which I would tear down and destroy were I able to do so, but this is beyond my power," Vaughan wrote. "They are part of my being and will perish only when I die or lose my memory."

When he remembered Fort Devens, this is what he saw: ". . . hundreds of stalwart young men in the uniform of their country coming into the wards of the hospital in groups of ten or more. They are placed on the cots until every bed is full yet others crowd in. Their faces soon wear a bluish cast; a distressing cough brings up the blood stained sputum. In the morning the dead bodies are stacked about the morgue like cord wood. This picture was painted on my memory cells at the division hospital, Camp Devens, in the fall of 1918, when the deadly influenza virus demonstrated the inferiority of human inventions in the destruction of human life."

It was shocking. Here was Vaughan, in the midst of the first war to use modern weapons, a war that was felling young men with machine guns and gas warfare, and yet it was all nothing compared to this illness.

The others, too, were traumatized. Cole was stunned by the hospital scene. As the men stumbled into the sick ward, "there were not enough nurses and the poor boys were putting themselves to bed on cots, which overflowed out of the wards on the porches," Cole said.

And then there was the autopsy room. It was hard to even get in, with stiff bodies piling up, blocking the doctors' way. "Owing to the rush and the great number of bodies coming into the morgue, they were placed on the floor without any order or system, and we had to step amongst them to get into the room where an autopsy was going on," Cole said.

But once they got there, even Welch, the imperturbable, the one the others looked to for courage and strength, was shaken. Somehow that was the worst of all.

Standing over the autopsy table, Welch opened the chest of the corpse of a young man, exposing his lungs. It was a terrible sight. "When the chest was opened and the blue swollen lungs were removed and opened, and Dr. Welch saw the wet, foamy surfaces with little real consolidation, he turned," Cole said. "This must be some new kind of infection," Welch said. "Or plague."

Welch "was quite excited and obviously very nervous," Cole said. "It was not surprising that the rest of us were disturbed, but it shocked me to find that the situation, momentarily at least, was too much even for Dr. Welch."

Cole, too, was shaken. It was, he said, "the only time I ever saw Dr. Welch really worried and disturbed."

By that time, the flu had spread beyond Fort Devens, beyond Boston, beyond the military. The entire state of Massachusetts was staggering from the virus.

Three days after Welch and his retinue visited Camp Devens, the state's health officials frantically wired for help, asking the U.S. Public Health Service to send doctors and nurses. The state's acting governor, Calvin Coolidge, sent a telegram to President

Woodrow Wilson, the mayor of Toronto, and the governors of Vermont, Maine, and Rhode Island, saying "our doctors and nurses are being thoroughly mobilized and worked to the limit." Many sick people, he added, "receive no attention whatsoever." As many as 50,000 in Massachusetts had the flu. That day, September 26, 1918, 123 Bostonians died of the flu and 33 succumbed to pneumonia.

But it was impossible to divert doctors and nurses to Massachusetts since, by then, the flu was everywhere and everyone needed help. The disease was moving quickly throughout military bases and towns and cities across the nation. Hundreds of towns, cities, and military installations were hit.

The result was devastation on a scale that is hard to imagine. Each incident, each military installation that was struck, each town or city, each remote village, had its own monstrous tale of death, helplessness, and social collapse.

The situation was so dire that the same day that Massachusetts pleaded for help, the Provost Marshal General of the U.S. Army made a shocking announcement. A draft call of 142,000 men was canceled. This despite the fact that soldiers were desperately needed in Europe. But he had little choice. The flu had spread far and wide. Twelve thousand Americans had died of the flu in September, and virtually every Army camp to which the recruits would report was under quarantine.

While Roy was administering to the dying in Fort Devens, while Welch was visiting the camp, looking on almost in disbelief at what the flu had wrought, the disease crept into Philadelphia.

Perhaps Philadelphia was ravaged early in the epidemic's course because the flu spread so easily from the city's Naval Yard. The flu first struck those Navy seamen on September 11, 1918, not long after it had arrived in Fort Devens. Or perhaps it was because the city was near two large Army camps, Fort Dix in New Jersey and Fort Meade in Maryland, and both of them were hit by the flu a few days later. Or perhaps the flu got its start in Philadelphia because the city had a huge Liberty Loan Drive parade, which drew a crowd of 200,000 on September 20. Or maybe it was all of these combined that gave the virus its foothold. But whatever the reason, Philadelphia was among the hardest hit of all American cities. And it was almost completely unprepared.

Few public officials anticipated the disaster and almost no members of the public did. The outbreak, in fact, was preceded by soothing words from medical authorities with a sort of band-played-on bravado. The *Journal of the American Medical Association* opined that medical authorities should not be alarmed by the flu's nickname, "the Spanish flu." That name, the journal wrote, "should not cause any greater importance to be attached to it, nor arouse any greater fear than would influenza without the new name." Moreover, the journal said, the flu "has already practically disappeared from the Allied troops."

Yet as the flu spread, the city did take a few precautions. On September 18, its health officials began a public campaign against coughing, spitting, and sneezing. Three days later, the city made influenza a reportable disease, which meant that records had to be kept of numbers of cases. On that same day, September 21, however, scientists reported good news—it seemed that the battle against influenza was won. The *Philadelphia Inquirer* wrote that researchers had found the cause of the flu—a bacterium called Pfeiffer's bacillus. As a consequence, the paper wrote, the finding has "armed the medical profession with absolute knowledge on which to base its campaign against this disease."

But by October 1, the city was under siege. In one day, 635 cases of the flu were reported to public health officials. That, however, was an underestimate. Doctors had become so overwhelmed caring for the sick that most cases went unreported and the true numbers will never be known. On October 3, the city closed all schools, churches, theaters, pool halls, and other places of amusement in a frantic attempt to slow the spread of the disease.

In the week that ended on October 5, as many as 2,600 were reported to have died in Philadelphia of the flu or its complications. The next week, the flu death reports reached more than 4,500. Hundreds of thousands were ill. Sick people arrived at teeming hospitals in limousines, horse carts, and pushcarts.

Within a month after the flu arrived in Philadelphia, nearly 11,000 people died from the disease. On one fateful day, October 10, 1918, 759 Philadelphia flu victims died.

"Visiting nurses often walked into scenes resembling those of the plague years of the fourteenth century," wrote historian Alfred W. Crosby. "They drew crowds of supplicants—or people shunned them for fear of the white gauze masks that they often wore. They could go out in the morning with a list of fifteen patients to see and end up seeing fifty. One nurse found a husband dead in the same room where his wife lay with newly born twins. It had been twenty-four hours since the death and the birth and the wife had had no food but an apple which happened to be within reach."

Undertakers were overwhelmed, observed Crosby. "On one occasion, the Society for Organizing Charity called 25 undertakers before finding one able and willing to bury a member of a poor family. In some cases, the dead were left in their homes for days. Private undertaking houses were overwhelmed and some were taking advantage of the situation by hiking prices as much as 600 percent. Complaints were made that cemetery officials were charging fifteen dollar burial fees and then making the bereaved dig graves for the dead themselves."

At the city morgue, bodies were piled three and four deep "in the corridors and in almost every room," Crosby said. They were "covered only with dirty and often bloodstained sheets. Most were unembalmed and without ice. Some were mortifying and emitting a nauseating stench. The doors of the building were left open, probably for circulation of air, and the Grand Guignol chaos was on view to anyone who cared to look in, including young children."

Philadelphia's nightmare was a prelude to an epidemic that roared throughout the world, bringing with it accumulating tales of horror. No place was safe, few families were spared. By the first week of October, the flu had spread to every part of the globe except for a few remote islands and Australia.

In Ottawa, Canada, the local newspaper reported that "street cars rattled down Bank Street with windows open and plenty of room inside. Schools, vaudeville theaters, movie palaces are dark; pool halls and bowling alleys, deserted."

In Cape Town, South Africa, there was such a shortage of coffins that bodies were buried in mass graves, wrapped in blankets.

Katherine Anne Porter, who had been a newspaper reporter in Denver, nearly died of the flu herself. Her fiancé was killed by the illness. She wrote a novella about her experience, *Pale Horse, Pale Rider*, a story told in almost dreamlike language and describing a nightmare: "All the theaters and nearly all of the shops and restaurants are closed, and the streets have been full of funerals all day and ambulances all night."

In Reading, England, a nurse wrote: "It happened so suddenly. In the morning we received an order to open up a new unit of flu and by night we'd moved into a converted convent. Almost before the desks were out the stretchers were in—60–80 to a classroom. We could hardly squeeze between the cots and oh, they were so sick! They came from a nearby airforce base . . . some had been lying unattended for days. They all had pneumonia. We knew those whose feet were black wouldn't live."

Buffalo Bill Cody lost his daughter-in-law and grandson. Writer Mary McCarthy was orphaned and sent to live with her uncle.

In France, John McCrae, a Canadian doctor assigned to the Medical Corps, had written the most famous poem about World War I, "In Flanders Fields." It is a paean to soldiers who died in battle: "In Flanders fields the poppies blow, between the crosses, row on row." McCrae himself died in the war, but not in battle. He was felled by pneumonia in 1918—which leading virologists say almost certainly was caused by influenza.

A doctor at the University of Missouri, D. G. Stine, wrote that from September 26 until December 6, 1918, 1,020 students got the flu. "I saw one patient die within 18 hours of this disease and 12 hours after being put to bed. I have seen a number of others menaced with death during the first 48 hours of the disease. The statement that influenza is uncomplicated is, I believe, erroneous," he wrote.

At Camp Sherman in Ohio, 13,161 men—about 40 percent of those at the camp—got the flu between September 27 and October 13, 1918. Of them, 1,101 died.

Army doctors tried every measure to stem the epidemic. They inoculated troops with vaccines made from body secretions taken from flu patients or from bacteria that they thought caused the disease. They made the men spray their throats each day and gargle with antiseptics or alcohol. They hung sheets between beds, and at one camp they even hung sheets in the centers of tables at mess halls. At Walter Reed Hospital, soldiers chewed tobacco each day, believing that it would ward off the flu.

Public health departments gave out gauze masks for people to wear in public. A New York doctor and collector of historical photographs, Dr. Stanley B. Burns, has a photograph in his archive of a minor league baseball game being played during the epidemic. It is a surreal image: The pitcher, the batter, every player, and every member of the crowd are wearing gauze masks.

In Tucson, Arizona, the board of health issued a ruling that "no person shall appear in any street, park, or place where any business is transacted, or in any other public place within the city of Tucson, without wearing a mask consisting of at least four thicknesses of butter cloth or at least seven thicknesses of ordinary gauze, covering both the nose and the mouth."

In Albuquerque, New Mexico, where schools were closed and movie theaters darkened, the local newspaper noted: "the ghost of fear walked everywhere, causing many a family circle to reunite because of the different members having nothing else to do but stay home."

Doctors gave out elixirs and vaccinated people against the flu, but to no avail. Crosby wondered about those flu vaccines. What was in them when no one knew what was causing the flu? He interviewed a doctor who had helped produce flu vaccines in 1918. The doctor, Crosby said, told him that the vaccines were just a soup made of blood and mucus of flu patients that had been filtered to get rid of large cells and debris. When they injected it into people's arms their arms became horribly sore. "So they thought it really worked."

Anecdotes spread. There was the story of four women who played bridge together one night. The next day, three were dead from the flu. There were tales of people who set off for work and died of influenza hours later.

Throughout the nation, families were ravaged. James D. H. Reefer of Kansas City wrote that when he was four and his brother was six, his thirty-year-old father and his twenty-seven-year-old mother got the flu and died within a few days of each other, unable to breathe as the flu destroyed the air sacs of their lungs. "Older members of the family later told me they 'simply drowned,'" Mr. Reefer said.

Minnie Lee Tratham McMullan was just two years old in 1918, living in Streator, Illinois. Her mother, her eleven-year-old brother, and her newborn baby sister died of the flu in that summer. An older brother got the flu and recovered, and so did Minnie, although at one time she was so ill that her family thought she was dead. "They rolled me out on the patio and put a sheet over me," she says. "Then they said they found out I was alive."

With his wife dead, Minnie's father was at a loss to care for Minnie, her older sister, and her two older brothers. The four children, aged two, four, seven, and ten, moved to the home of one relative and then the next.

Years later, Minnie McMullan went to the cemetery in Streator and spoke to the caretaker, who told her that the dead were lined up along the road during that terrible time. "There were so many dying that they didn't have people to dig graves to put them in," she says.

But McMullan, the last member of her family who was alive in 1918, has no memory of the epidemic. All she knows is what her relatives told her, and they did not like to speak of the deaths and dying. "I'm glad I don't remember," she says.

At the same time, volunteers, largely women, bravely stepped forward to minister to the sick. In El Paso, Texas, where poor Mexicans were dying at alarming rates, a twenty-eight-room school, the Aoy School, was converted into a hospital for influenza patients, mostly Mexicans. The local newspaper described the scene on October 19: "Fifty-one Mexican men, women, and babies lay gasping in the improvised wards of Aoy School last night." They were "brought from the squalor of homes in the Mexican quarter of town, many of them in the last stages of pneumonia, all of them suffering from the lack of proper medical attention and comforts, the patients were transferred from the depths of poverty to the comparative comfort and care of a hospital equal in almost any respect to any in the city."

People from all over the city volunteered at the Aoy School, providing food and clothing, driving the sick to the hospital in their own cars. Women helped out as cooks, clerks, drivers, and nurse's aides. One wrote: "I am so glad I can help. I have not had a nurse's aide course, nor, in fact, any training. I probably have no qualifications for nursing except for my desire to relieve some of the suffering."

Perhaps none but a gifted novelist can tell what death from the 1918 flu looked like, how the stricken person appeared in those last hours of life when the horrors of the illness are fully unfurled. One of the few who attempted this was Thomas Wolfe. He was a student at the University of North Carolina in 1918 when he got a telegram from home, summoning him to return immediately. His brother, Benjamin Harrison Wolfe, was ill with the flu. He tells the thinly fictionalized story in Chapter 35 of his novel, *Look Homeward, Angel*.

Wolfe came home to a deathwatch. His brother was lying in a sickroom upstairs while his family waited for what they feared was inevitable. Wolfe went upstairs to the "gray, shaded light" of the room where Ben lay. And he saw, "in that moment of searing recognition," that his beloved twenty-six-year-old brother was dying.

"Ben's long thin body lay three-quarters covered by the bedding; its gaunt outline was bitterly twisted below the covers, in an attitude of struggle and torture. It seemed not to belong to him, it was somehow distorted and detached as if it belonged to a beheaded criminal. And the sallow yellow of his face had turned gray; out of this granite tint of death, lit by two red flags of fever, the stiff black furze of a three-day beard was growing. The beard was somehow horrible; it recalled the corrupt vitality of hair, which can grow from a rotting corpse. And Ben's thin lips were lifted, in a constant grimace of torture and strangulation, above his white somehow dead-looking teeth, as inch by inch he gasped, a thread of air into his lungs.

"And the sound of this gasping—loud, hoarse, rapid, unbelievable, filling the room, and orchestrating every moment in it—gave to the scene its final note of horror."

The next day, Ben grew delirious. "By four o'clock it was apparent that death was near," Wolfe wrote. "Ben had brief periods of consciousness, unconsciousness, and delirium—but most of the time he was delirious. His breathing was easier, he hummed snatches of popular songs, some old and forgotten, called up now from the lost and secret adyts of his childhood; but always he returned, in his quiet humming, to a popular song of war-time—cheap, sentimental, but now tragically moving: 'Just a Baby's Prayer at Twilight.'"

And then Ben sank into unconsciousness. "His eyes were almost closed; their gray flicker was dulled, coated with the sheen of insensibility and death. He lay quietly upon his back, very straight, without sign of pain, and with a curious upturned thrust of his sharp thin face. His mouth was firmly shut."

Wolfe stayed with Ben that night, fervently praying even though he had thought he did not believe in God or prayer. "'Whoever You Are, be good to Ben to-night. Show him the way . . . Whoever You Are, be good to Ben to-night. Show him the way . . .' He lost count of the minutes, the hours: he heard only the feeble rattle of dying breath, and his wild synchronic prayer."

Wolfe fell asleep, then woke suddenly, calling his family in with a certain knowledge that the end was nigh. Ben quieted, lay still. "The body appeared to grow rigid before them." Then, in a last gasp, "Ben drew upon the air in a long and powerful respiration; his gray eyes opened. Filled with a terrible vision of all life in the one moment, he seemed to rise forward bodily from his pillows without support—a flame, a light, a glory." And so, Wolfe wrote, Ben "passed instantly, scornful and unafraid, as he had lived, into the shades of death."

Nothing could be done for Thomas Wolfe's brother, Ben. No one knew how to treat the flu. There was no medicine to quell the raging fevers, no way to get oxygen into sodden lungs. There was no way to prolong life and no way to soothe the dying. Treatment was what doctors called palliative—give the patient food, fresh air, if possible, and TLC. Those optimistic tales, told when the flu struck Philadelphia, that a bacterium that caused the flu had been isolated, turned out to be untrue. Yes, the bacterium was found. But no, it did not seem to lead to a treatment or a vaccine. The cause of the disease remained a mystery. In 1918, the widely proclaimed discovery of Pfeiffer's bacillus turned out to be a false trail. The influenza virus was beyond anyone's grasp.

It was not just that the epidemic struck during wartime, when the nation was distracted by the horrors of battle. It was also that the epidemic came before scientists had any idea of how to isolate an influenza virus and discern its secrets. It was a time when the germ theory of disease was known, and when scientists had discovered that there was such a thing as a virus. But no one had ever seen a virus—electron microscopes had not been invented and viruses are far too small to be seen with ordinary microscopes. And no one understood what viruses were, since DNA and RNA, the genetic material of viruses and the clues to their destructiveness, had not yet been discovered.

Even today, with the exquisite advances of molecular biology and the pharmaceutical industry, viral diseases—and influenza in particular—are largely untreatable. It's not that molecular biologists are ignorant about the inner workings of influenza viruses. They have known for decades that the simple influenza virus has only eight genes, each made of the material RNA, and that the viruses die in hours if left alone with no cells to infect. They know what the flu viruses look like—under an electron microscope, they are little balls or egg-shaped particles, although sometimes they form long filaments. They know how they are put together—flu virus particles are wrapped in a slippery fatty membrane, held in place by a protein scaffolding underneath. They know how the viruses burrow into a cell and burst out again by using hundreds of sharp protein shards that poke out of the virus's membrane. They even know why human influenza viruses infect only cells of the lungs—those are the only human cells with an enzyme that the virus needs to split one of its proteins during the manufacturing of new virus particles.

But what they do not know is how to make a medicine that is the equivalent of a penicillin for the flu. The best way to combat an influenza pandemic is with vaccines—if manufacturers have enough advance notice of a new flu strain to make enough vaccine to go around. Manufacturers could, if they understood what made the 1918 flu so deadly, stockpile vaccines sufficient to protect the population if that flu or one like it ever came again. That, however, would require knowing what the 1918 flu looked like. Yet the last victims of the flu died in 1918, taking the virus with them.

Under ordinary circumstances, that would be the end of the story. The flu lived in the soft tissues of the lungs, and the lungs decay almost immediately after death. The virus, in fact, should be gone long before a corpse's lungs decay.

But nothing about the 1918 flu was ordinary. And perhaps the most extraordinary story yet comes nearly a century later, when three people among the millions who died of the flu are turning out to contain, within their miraculously preserved lung tissue, a sort of Rosetta stone for the killer flu virus. Unknown to anyone when these three people suddenly died, they alone would hold the clues to protect the world in the twenty-first century.

The first of the three to become ill with influenza was an Army private, Roscoe Vaughan, just twenty-one years old in September 1918. Like every soldier, he must have been frightened and full of bravado. He must have expected to be in battles and have hoped he could be strong and true. As he arrived at Camp Jackson, seven miles east of Columbia, South Carolina, he joined a contingent of more than 43,000 young men being trained in artillery before they went overseas. They practiced maneuvers on the dunes, struggling over the loose, drifting sand, squinting in the bright glare of the South Carolina sun. Roscoe Vaughan was among them, fit and healthy. He surely thought that the adventure of his life was about to begin. In a way, it was.

Private Vaughan was unlucky enough to have come to a camp staggering under a sudden onslaught of influenza. The soldiers were easy prey for the flu, and the hospital, built high on a sandy hill, was flooded with sick young men. In August, 4,807 men were admitted as patients. In September, 9,598. One young doctor there, James Howard Park, Jr., said he literally watched men drop dead as they walked across the paths at the camp. One day he himself tagged thirty corpses.

Private Vaughan's medical records show that he became ill during the third week of September, feeling achy and feverish. It did not take long for the virus to do its work. On September 19, he reported to sick call, stricken with the flu. At 6:30 in the morning of September 26, he died, gasping for breath.

At 2 p.m., an Army doctor, Captain K. P. Hegeforth, arrived to do an autopsy. Private Vaughan, he wrote, was a "fairly well developed, well nourished man measuring five feet ten inches." He was chubby, with a "moderately thick layer of subcutaneous fat." But he was fit, with "muscles in good condition."

Private Vaughan had about 300 cc, or 1¼ cups, of clear fluid in his chest cavity. Over the entire surface of his left lung were small seepages of blood, ranging in size from a pinhead to a dime or larger. The air sacs were clogged with fluid.

Captain Hegeforth cut off a slice of one of Private Vaughan's sodden lungs for examination, impregnating the slice of lung with formaldehyde to preserve it, embedding it in a chunk of candle wax about the size of a thumbnail. Then he sent it to Washington, where it was stored in a small brown box on a shelf of a vast government warehouse.

While Private Vaughan was dying in South Carolina, Private James Downs, a thirty-year-old soldier at Camp Upton in New York, was also feeling ill. He had arrived at the camp, about sixty-five miles east of New York City, to prepare to go overseas to fight. It was a setting not very different from the one Private Vaughan found himself in—an embarkation camp on a flat, sandy stretch of land, this time between Long Island Sound and the Atlantic Ocean, dotted with pine trees and low sandy bushes. But the camp, built a year earlier, was jammed with 33,000 soldiers who struggled to drive over the nearly impassable roads.

In September 1918, the hospital was suddenly flooded with sick men—one out of every ten men in the camp was admitted to the hospital. Private Downs was among them. His records show he entered the base hospital on September 23. His face was flushed, he was delirious, and his temperature was 104 degrees.

The next day, he remained delirious, with the same high temperature, but his skin was turning dark from lack of oxygen. At 4:30 a.m. on September 26, three days after he was admitted to the hospital, Private Downs died of influenza, just two hours before Private Vaughan succumbed.

That same day, a Captain McBurney did an autopsy on Private Downs. He wrote that Downs was six feet tall, weighing 140 pounds. He had "no external signs of disease or injury."

The injury, of course, was internal—in Private Downs's lungs. They were filled with fluid, and a "bloody froth" exuded from them, Captain McBurney wrote. The doctor cut a small slice of James Downs's lung, soaked it in formaldehyde, encased it in wax, and sent it to the same Washington warehouse that stored the scrap of Private Vaughan's lung.

And there the lung specimens from Private Vaughan and Private Downs remained, for nearly eighty years, secreted among pathology tissue from millions of people who died of diseases both common and rare, in the Armed Forces Institute of Pathology's archives. The archives had begun in the Civil War, created by an executive order by President Abraham Lincoln. Since then, military doctors had been sending in thousands of pathology specimens a year, with as many as 50,000 a year sent in more recently. The number of tissue samples stored in the warehouse had swelled to about 3 to 4 million.

Several times in the past century, the warehouse specimens were moved as the ever-growing collection required more and more storage space. But the tiny paraffin cubes encasing lung cells from the autopsies of Roscoe Vaughan and James Downs remained in their boxes, of no interest to anyone until, at the end of the twentieth century, they were rediscovered by molecular biologists who thought it just might be possible to resurrect the flu virus from those ancient slices of lung.

Two months after Private Vaughan's death, influenza came to Teller Lutheran mission (now called Brevig), on the flat frozen tundra of the Seward Peninsula of Alaska. It was an isolated village of eighty people, ninety miles by dog sled from Nome, the nearest town, perched on the edge of a gray and chilly sea. Among the residents was an obese woman who, like the rest, lived in a soot-filled igloo with a window made of seal gut.

On the last Saturday in November, two visitors from Nome attended a crowded, standing-room-only service in the tiny chapel run by the missionaries. The visitors reported that there was much sickness in the city, but no one was unduly alarmed. The Eskimos treated the visitors to their traditional hospitality—a feast for the entire village of reindeer meat, hotcakes, blueberries in seal oil, and tea.

Two days later, on Monday, the first villagers fell ill with the flu. On Tuesday, the first person died, a woman named Mrs. Neelak. The minister set off for Teller, a village fourteen miles from the mission, seeking help. He returned, only to report that people in Teller were being decimated by the same illness.

One after another, the Eskimos died—seventy-two in all. One igloo had twenty-five dead bodies, frozen in the Arctic cold. Starving dogs had broken into another igloo and ripped apart a collection of corpses, leaving a gory mess of bones. Another igloo looked at first like the site of utter devastation. The seal-gut window had broken and snow had drifted in. The fire was out, the bone-chilling cold had penetrated the small space. And as rescuers peeked inside, they saw only a pile of corpses. Then, suddenly, three terrified children appeared from under deerskins and started shrieking. They had survived somehow on oatmeal, surrounded by the bodies of their family.

In the end, a three-week flu epidemic had left only five adults in the village; forty-six children were orphaned. Clara Fosso, the missionary's wife and one of the few adults who did not get the flu, wrote a letter to the Eskimos years later, still mourning the tragedy: "There was a spiritual revival among the Eskimos at the Mission on the last Sunday in November 1918, before the influenza disaster fell upon us. The whole settlement of Eskimos had crowded into the new school room for worship. We felt the spirit of the Lord among us, as the communicants stood at the altar and later met in prayer; many confessed to their faith. We were deeply moved. This was the last time we were gathered together. By the following Sunday most members had gone to a more beautiful service with their Savior. You, who are the sons and daughters of these children of God, may remember that many of them died testifying to their Lord and singing the hymn that we had shared on that last Sunday, 'I Can Hear My Savior Calling.'"

Still in shock from the tragedy, the few able-bodied men that could be found in the nearby villages were left to bury the dead, an ordeal in the harsh Alaskan winter. The villages were so far north that the ground was permanently frozen. In order to dig into the rock-hard ground at Teller mission, miners pumped hot steam into the frozen earth, thawing it enough to dig a trench. They buried the flu victims in a common grave, marking the site with two large crosses, one at each end of the trench.

Among the dead was the obese woman, whose frozen body lay six feet underground in the mass grave. And there it remained for seventy years.

# 2

# A HISTORY OF DISEASE AND DEATH

The young man grew up in Athens surrounded by luxury and showered with privilege. His father owned gold mines, so money was never a problem. A brilliant scholar, the man indulged himself by spending his days discussing philosophy with one of the great masters of his age. It was a leisurely life of the mind, unsullied by worldly concerns.

Until, one day, a plague descended upon the city.

The sickness came in the year 431 B.C. The citizens were used to illnesses and deaths but nothing had prepared them for this tragedy. For more than a year, the epidemic raged, laying waste to every structure of the carefully built society, creating a cataclysm that shook the confidence of scientists and doctors, changing history. It seemed truly a work of wrathful gods. The young man, Thucydides, was its chronicler.

The symptoms were frightening. Strong and healthy young people would suddenly be afflicted with "violent heats in the head, and redness and inflammation in the eyes, the inward parts, such as the throat or tongue, becoming bloody and emitting an unnatural and fetid breath," Thucydides wrote. The hapless victims sneezed, their voices became hoarse. Their chests hurt as they produced "a hard cough." Some suffered agonies of abdominal cramps. "Discharges of bile of every kind named by physicians soon ensued, accompanied by great distress." People would retch and heave as their intestines went into spasms.

The plague's victims were so hot they felt on fire and they burned with thirst. They yearned to plunge into cold water, and some did, hurling themselves into rain tanks, trying to slake their "agonies of an unquenchable thirst."

Sick and dying citizens turned to their doctors, pleading for relief. But to no avail. No medicine, no salve could ease the suffering. Even worse, the doctors themselves began falling ill from exposure to the legions of the sick.

Terrified citizens turned from medicine to religion, crowding into temples and praying to the gods to grant them relief. But it was no use, Thucydides explained. "Supplications in the temple, divinations, and so forth were found equally futile, until the overwhelming nature of the disaster at last put a stop to them altogether."

Thucydides described scenes out of nightmares: "The bodies of dying men lay upon one another, and half-dead creatures reeled about the streets and gathered round all the fountains in their longing for water." So many died that burial rituals went by the wayside and "they buried the bodies as best they could."

No one knew what to do or where to turn for help. A sniffle or a headache could be the first sign of doom, and there was no way to stave off the sickness once it took hold. Those who began to feel ill would despair and that despair itself "into which they instantly fell took away their power of resistance, and left them a much easier prey to the disorder."

It was normal—expected, in fact—for the healthy to aid the sick. Those who tried, however, could catch the disease themselves. Athenians had to decide: Should they nurse their friends and relatives and risk becoming ill, "dying like sheep"? Or should they turn coldly away and attempt to save themselves?

The answer, for the frightened Athenians, was obvious. People began holing up in their homes, afraid to visit friends, relatives, or neighbors. The sick began to perish from neglect.

The order of a civil society quickly disintegrated. The epidemics, Thucydides relates, brought forth a wild recklessness, a "lawless extravagance which owed its origins to the plague."

"Men now coolly ventured on what they had formerly done in a corner," he said. Athenians could not forget the sight of rich men dying suddenly and those who had had nothing seizing the rich men's property. What was the use of saving or living frugally, many asked, when death could come at any moment and the poor could descend like vultures? "So they resolved to spend quickly and enjoy themselves, regarding their lives and riches as alike things of a day."

Traditional notions of honor were abandoned, and, instead, "it was settled that present enjoyment, and all that contributed to it, was both honorable and useful." Lawlessness took hold. "Fear of gods or law of man there was not to restrain them," Thucydides wrote. Why worship gods, many asked, when "they judged it to be just the same whether they worshipped them or not, as they saw all alike perishing"? And why obey laws when "no one expected to live to be brought to trial for his offenses"?

The city was never the same after that plague. In fact, Thucydides implied that the plague was a reason why Athens failed in its plans to defeat Sparta and the Peloponnesian League.

Thucydides' story marked the start of the age of chronicled plagues. To this day, no one knows the source of the illness that struck Athens and no one can convincingly rule out competing theories. Influenza? Toxic shock? What remains is the horror.

Indeed, the story of humanity before the twentieth century is the story of an often losing battle between people and disease.

Periodically, arising from the swarm of microbes that traveled with people from time immemorial, came epidemics that nearly wiped out populations, almost overnight. They always were unexpected and they always were inexplicable. Was it the wrath of God? Was it the miasma, the bad air, that made the sickness spread? No one could say, and the wisest men were helpless under the disease's onslaught.

Until the twentieth century, infectious diseases were so common and so untreatable that it was difficult for populations even to maintain their numbers in the face of epidemics. There were lulls, but each lull seemed to be followed by the fury of another terrifying infectious disease.

Diseases like tuberculosis were endemic in crowded urban areas and mortality rates from infectious diseases were so high in places like London that it was not until 1900—the first time since cities came into existence five thousand years earlier—that large cities could maintain their populations without constant influxes of immigrants.

The worst were plagues that changed the course of history and spelled doom for societies. They even changed human evolution. Survivors of plagues were the genetically lucky ones who had inherited a resistance to the disease-causing organisms. Even in the most extreme plagues, there are resistant people who either do not become infected, no matter how many times they are exposed to the sickness, or who get only a mild disease and recover. When everyone else is dying, the resistant people will be the ones who remain to propagate. Their genes will begin to predominate. And those who were genetically susceptible to the devastating illnesses would lose out in the great Darwinian struggle.

In the epic sea of plagues, one epidemic stands out. It came a millennium after Thucydides wrote about the plague of Athens and it swept most of the world, leaving death and devastation in its path.

Medical historians believe the sickness began in China in 1331. Along with a civil war, it halved the Chinese population.

From there, the plague moved along trade routes of Asia and arrived in the Crimea fifteen years later, in 1346. Then it entered Europe, North Africa, and the Middle East. It disrupted society in ways eerily reminiscent of the Athens plague so long before. It emptied streets and public places like the flu epidemic that followed it. And its very name became emblematic of the horrors of epidemics. It was known as the Black Death.

At the time the illness was as mysterious as the plague of Athens but now it is known that the Black Death bacteria, *Yersinia pestis*, were spread by fleas that lived on black rats. The rats, in turn, moved from port to port on ships, taking the illness with them. The fleas would bite people, infecting them with the bacteria.

The plague would not have been so overwhelming if it could only spread through flea bites. It turned out that once the bacteria began infecting people, they found another way of spreading. They would infect the lungs and cause a pneumonia, whereupon sick people could infect the healthy simply by coughing or sneezing. With that mode of infection, there was no stopping the epidemic.

The Black Death came at a time when Europe had been relatively free of disease for three centuries and when its population, tamped down for hundreds of years by illnesses, had tripled. Europeans had grown prosperous and optimistic. Then they were struck with a catastrophe. In the few short years from 1347 to 1351 the sickness killed at least a third of the European population.

Chroniclers of the plague plaintively described its devastation. In Siena, Italy, where half the population of 60,000 died, Agnolo di Tura wrote: "It is impossible for the human tongue to recount the awful truth. Indeed, one who did not see such horribleness can be called blessed. And the victims died almost immediately. They would swell beneath the armpits and in their groins, and fall over while talking. Father abandoned child, wife husband, one brother another . . . And in many places in Siena great pits were dug and piled deep with the multitude of dead . . . And as soon as those ditches were filled, more were dug . . . And so many died that all believed it was the end of the world."

The scene in Florence resembled Katherine Anne Porter's account of Denver during the 1918 flu. Between 45 and 75 percent of the inhabitants of the Italian city were killed by the Black Death and the streets were bereft of crowds. Only carts and wagons rattled down the roads, scooping up bodies of the dead.

The dead bodies of plague victims were easy to find. Giovanni Boccaccio wrote in his *Decameron* that people, afraid of contamination by the rotting corpses, would drag the dead outside their houses and leave them in front of their doors to be picked up, like so much garbage.

Funeral processions often included more corpses than priests had anticipated: "And times without number it happened that two priests would be on their way to bury someone, holding a cross before them, only to find that bearers carrying three or four additional biers would fall in behind them; so that whereas the priests had thought they had only one burial to attend to, they in fact had six or seven or even more."

But, Boccaccio added, people quickly hardened their hearts to the dead. "There were no tears of mourners to honor the dead," he wrote. "In fact, no more respect was accorded to dead people than would nowadays be accorded to dead goats."

People were changed by the epidemic. Boccaccio told of two extremes. One group of terrified citizens hid out from society, locking themselves inside of homes where no one had gotten sick. There, they "settled down to a peaceable existence, consuming modest quantities of delicate foods and precious wines and avoiding all excesses. They refrained from speaking to outsiders, refused to receive news of the dead or the sick, and entertained themselves with music and whatever other amusements they were able to devise."

At the other extreme were those who fled to lives of wild abandon. They "maintained that an infallible way of warding off this appalling evil was to drink heavily, enjoy life to the full, go round singing and merrymaking, and gratify all of one's cravings whenever the opportunity offered, and shrug the whole thing off as one enormous joke."

Some groups of revelers went from one tavern to another, drinking all day and throughout the night. Others would hole up in private homes that they requisitioned. There they would fortify themselves with drink, insisting that "conversation was restricted to subjects that were pleasant or entertaining."

No one was left to enforce laws of society or religion, Boccaccio wrote. For "all respect for the laws of God and man had virtually broken down and been extinguished in our city."

Many fled, abandoning "their city, their homes, their relatives, their estate and their belongings." But the plague had also ravished the countryside. Peasants stopped tilling their fields and neglected their animals, which roamed the fields. Like the city residents, those who lived in small towns and on farms "behaved as if each day was to be their last," and "tried in every way to squander the assets in their possession."

The Black Death finally went away, perhaps because the bacteria had already infected all who were susceptible. But other deadly epidemics continued to sweep the world, and even old diseases, like cholera, sometimes took on a fierce vigor and caused plagues as terrible as any that had afflicted humanity.

William Sproat thought little of the attack of diarrhea that struck him on Saturday, October 23, 1831. He recovered immediately and simply put it out of his mind. When dinnertime arrived, he ate a mutton chop and then walked from his home in Sunderland in Britain's Durham County to the boat on a nearby river where he worked as a keelman.

The agonizing stomach cramps began when he arrived at his boat. Sproat doubled over in pain, and a watery, white-speckled diarrhea poured from his bowels, seemingly gallons of it. Each attack was heralded by violent intestinal spasms. He began vomiting. Sproat barely made it home, where he crawled into bed, shivering with fever, writhing in pain.

That night, Sproat curled in his bed, unable to sleep as the symptoms continued unabated. The next day, his doctor arrived. Sproat, he wrote in his notes, was "evidently sinking; pulse almost imperceptible, and extremities cold, skin dry, eyes sunk, lips blue, features shrunk, he spoke in whispers, violent vomiting and purging, cramps of his calves and legs, and complete prostration of strength."

On Wednesday morning, Sproat's doctor wrote an even more dire report: "Pulse scarcely beating under the finger, countenance quite shrunken, lips dark blue." Sproat died at noon, the first known victim of the great cholera epidemic to follow.

His son was next, dying of the awful illness a few days later. His granddaughter too became ill, with the identical symptoms, but she recovered. Others in Sproat's town got the sickness. Within weeks, the disease had spread throughout the land, leaving a swath of dead.

It was a sickness notable for its sudden onset, terrifying and unpredictable spread, and horrifying mortality. Untreated, cholera kills 40 to 60 percent of its victims. Today it is treated with fluid replenishment, including intravenously. But no one in the early nineteenth century knew how to control the disease or even what caused it or how it spread. When Sproat became ill, the disease was a complete mystery.

While cholera itself was an ancient illness, the disease this time seemed particularly virulent. It had first been noted by British newspapers in 1818 when it appeared in India. Reporters described it as a new disease, calling it cholera morbus, that was burning through Calcutta and killing British troops fighting there. In the winter of 1818–19, cholera killed 3,000 of a British army of 10,000 in India that was led by the Marquis of Hastings. One London doctor who traveled with the army to India wrote of his trepidation. "To a young man of twenty-five, what a fearful responsibility was the charge of a European regiment, with 2,000 native camp followers and the cholera among them coming as an epidemic. Never before in my life had I seen anything as dreadful as this disorder."

The disease stayed clear of England, however, until 1831, when William Sproat fell ill. The epidemic that followed was the first of a wave of six deadly cholera epidemics that spread across the world, frightening and killing multitudes. British historian R. J. Morris writes that in Britain "the approach and arrival of these epidemics, especially that of 1832, created a crisis atmosphere in the country quite unlike that produced by any other threat apart from a foreign invasion." At least 140,000 died in Britain. Morris adds that a "normally calm" journal, the *Quarterly Review,* described cholera as "one of the most terrible pestilences which ever have devastated the earth."

Sproat was the typical cholera victim. The disease is caused by a lozenge-shaped bacterium, *Vibrio cholera,* that is usually transmitted when someone drinks contaminated water. But the bacteria also can be transmitted by food or by flies, on blankets and on clothing. The bacteria can survive for two to five days in foods like meat, milk, and cheese, but it can live for as long as sixteen days on apples.

Cholera bacteria release a powerful toxin into the intestines that forces cells to pump water and salts from blood and tissues. The result is massive watery diarrhea that is dotted with mucus and skin cells, the so-called rice-water stool.

And as the bacteria draw fluids from the victim's body, the vital signs start to plummet. Within an hour after the symptoms begin, a healthy person's blood pressure can drop precipitously. The dehydration and loss of salts creates excruciating cramps. Death can follow two to three hours later, but more often occurs within eighteen hours to several days after the first attack of diarrhea.

One Edinburgh doctor, George Bell, who had witnessed cholera in India, wrote about the disease for his British colleagues: "The eyes surrounded by a dark circle are completely sunk in the sockets, for the whole countenance is collapsed, the skin is livid. The surface [of the skin] is now generally covered with cold sweat, the nails are blue, and the skin of the hands and feet are corrugated as if they had been long steeped in water . . . The voice is hollow and unnatural. If the case be accompanied by spasms, the suffering of the patient is much aggravated, and is sometimes excruciating."

As the cholera epidemics raged in Great Britain, the newspapers kept a running toll of the cases and deaths, which had "the depressing effect of the tolling of a funeral bell," Morris writes.

Some small villages were nearly destroyed. Of Bilston, an iron-and-coal-mining town in Staffordshire, observers wrote: "To describe the consternation of the people is impossible. Many factories and workshops are closed; business is completely at a stand; women seen in a state of distraction running in all directions for medical help for dying husbands, husbands for their wives, and children for their parents; the hearse carrying the dead to the grave without intermission either by day or night. Those inhabitants who possessed the means of quitting their homes, flying to some purer atmosphere; those who remained seeing nothing before them but disaster and death."

In other countries, there was a mass exodus from large cities, as all who could fled the epidemic. In Moscow, 50,000 left, and in Paris, 700 people deserted the city each day as cholera raged.

Religious leaders in Britain tended to see the epidemic as a manifestation of the wrath of God and a call for the people to turn to their faith for salvation. Some said God sent the epidemic to counter the "pride" and the "impotent boastings of modern science." In one English Methodist chapel, an observer wrote: "Meetings were held every evening of the week, and the chapel was filled; many groaned under a sense of their sin and guilt, sought mercy with cries and tears, and would hardly leave the chapel at a late hour."

When the 1832 cholera epidemic in Britain waned, the weary population just wanted to forget it. People had grown tired of reading and discussing an epidemic that had finally dissipated.

The leading medical periodical, the *Edinburgh Medical and Surgical Journal*, announced it would stop reviewing books on cholera, explaining that its editors reached that decision because of "the multitude of books which have recently issued from the press on the subject of cholera, and our determination no longer to try the patience of our readers." Cholera disappeared as a subject in newspapers and magazines.

From Sproat to amnesia. It seems unreal. How could something so terrible as an infectious disease that kills half its victims, and gruesomely, be so quickly relegated to the dustbins of history?

Morris ventures a guess, actually several guesses. First, he says, people had expected that the terrifying epidemic would cause a massive social disruption. That did not, in the end, occur. And second, "there were no clear 'lessons.'" The main effect of the epidemic on British society, he claims, was to spur the efforts of a group of public health researchers who were to prevail in implementing dramatically effective measures—like cleaning up the water supply—in the coming decades.

The cholera epidemic was a turning point marking the last time the disease would rage without simple precautions of public health. John Snow, a British doctor, had watched helplessly as cholera ravaged the population in the 1830s and had begun to question the prevailing hypothesis which maintained that the disease was spread by a "miasma"—odoriferous gases given off by rotting vegetables or flesh. How could it be spread by gases, he asked, when it was a disease of the intestines, not the lungs? Snow began to suspect that cholera was spread in water.

In August 1854, cholera roared into London's Soho district, killing ninety-three people. Snow, who by then had become London's leading anesthetist, decided to investigate, conducting a famous experiment that finally broke the back of the disease. He graphed the deaths from cholera and noticed that they seemed to occur in people who drank from one of several public wells. The water in that well, he proposed, must be the culprit. To test his hypothesis, Snow removed the pump handle from the suspect well and the cholera epidemic came to an abrupt end.

In 1883, Robert Koch, one of the founders of the germ theory of disease, went to Egypt to uncover the cause of a cholera epidemic that seemed to be moving toward Europe. He arrived just after a colleague of the great Louis Pasteur, Pierre Emil Roux, who was trying but failing to isolate the cholera-causing microorganism. He was trying to grow the bacteria the way the master had taught him—in broth, which tended to be contaminated with other microorganisms. Koch had a better laboratory method, growing the bacteria on the springy surface of agar, where he could see and discard contaminants. Not only did he find the microorganism, the comma-shaped *Vibrio cholerae*, in Egyptian patients, the next year he showed that the bacteria lived in human intestines and that they were transmitted in water. He repeated the work in Calcutta and reported his victory to the German government, which hailed him as a hero.

Not everyone was won over, however. One Munich hygienist, Max von Pettenkofer, insisted that the miasma theory was correct and, to prove it, he asked Koch for a flask of broth brimming with cholera bacteria. He drank it down. He wrote back to Koch jubilantly: "Herr Doctor Pettenkofer has now drunk the entire contents and is happy to be able to inform Herr Doctor Professor Koch that he remains in his usual good health." To which one modern commentator, Roy Porter, a historian of medicine at the

Wellcome Institute for the History of Science in London, wryly notes, "Pettenkofer must have been fortunate enough to possess the high stomach acidity which sometimes neutralizes the vibrios." But while Pettenkofer is a curious footnote in history, Koch triumphed. Soon, Porter writes, Koch was "burdened with success." As a consequence, "his research declined, and so to offset that he turned oracle."

The victory over cholera was only a beginning. With the growing and profound knowledge that many diseases are caused by microscopic organisms and that the spread of disease can be prevented, the Western world was transformed. It took years for the change to be complete, but the result was a vigorous public health movement that emphasized simple but powerful measures like cleaning up water supplies and teaching people what now seem to be basic lessons of health and hygiene—keep flies away from food, wash your hands before handling food, give your babies milk, not beer, quarantine the sick. The results were dramatic. In large areas of the world, many of the killer diseases seemed tamed, or even vanquished, and deadly epidemics seemed to be relics of the past.

In England and Wales, for example, deaths from tuberculosis, a major killer, had dropped by 57 percent by the turn of the century. The mortality rate from measles had plummeted by 82 percent. Even the mysterious yellow fever was vanquished and its cause—the first human virus ever discovered—was elucidated.

By the dawn of the twentieth century, for the first time since cities had come into existence 5,000 years before, infectious diseases were staunched to such an extent that cities were able to remain stable, and even grow, without depending on a constant stream of migrants from the countryside. It was a remarkable change.

The medical miracles altered the way soldiers fought in World War I. Scientists had discovered that one of the worst killers of soldiers—typhus—was spread by lice. As a consequence, doctors insisted the troops fighting in World War I undergo rigorous delousing procedures. It was this successful effort to keep typhus at bay, in conjunction with widespread vaccination of troops against other common illnesses, that allowed troops to fight in such close quarters in trenches. In fact, the only epidemic disease that plagued the troops during the early years of World War I was syphilis.

As the second decade of the twentieth century wound to a close, the memory of a world stalked by infectious disease had dimmed. People had become complacent, almost smug, about disease and death. It was a time when death had nearly lost its sting, an era when the miracles of medicine were portrayed as almost a new religion. And it was a time when death became separate from everyday life. *The Ladies' Home Journal* proudly declared that the parlor, where the dead had been laid out for viewing, was now to be called the "living room," a room for the living, not the dead.

Then the influenza epidemic arrived. Unlike the plague of Athens, unlike the Black Death, unlike even the cholera epidemic that felled William Sproat and the other cholera epidemics to come in that century, the flu epidemic had no chronicler.

Dr. Victor C. Vaughan seated himself comfortably before the warmth of his fireplace, took pen in hand, and settled down to write his memoirs. The sixty-seven-year-old man had retired after a long and august career, having held the highest honors of American medicine. He knew he had a powerful story to tell.

The influenza epidemic was part of his story, of course. At the end of his professional life, in October 1918, Vaughan—as has been noted—had visited Fort Devens near Boston and had seen for himself the start of the epidemic. As he sat down to write, just a few years had passed since the portly doctor, graying at the temples, had made the rounds of the hospital wards at that Army camp. He had solemnly walked among the young soldiers, laid out on hospital beds and dying, row after row of them, their sheets stained with blood, their mouths frothy with blood-tinged sputum, their skin dark from lack of oxygen as they gasped for breath. He had seen the piles of corpses. He had been awestruck by the utter destructive power of this new disease.

Vaughan came to realize that Fort Devens was just the beginning. He had watched as the influenza epidemic spread to the farthest corners of the world, killing millions, crippling armies, until, as mysteriously as it came, it disappeared.

In Vaughan's 464-page memoir, published in 1926, readers might have expected to learn about the 1918 flu and Vaughan's still-raw memories of the plague years. If so, they would have been disappointed. Instead of reflecting on the epidemic, Vaughan chose not to dwell on it, taking note of Fort Devens in a single paragraph where he described the scene there as one of "the gruesome pictures exhibited by the revolving memory cylinders in the brain of an old epidemiologist as he sits in front of the burning logs on the hearth."

When Vaughan wrote about the war, in which influenza killed more men than combat, he devoted a mere two sentences to the epidemic: "I am not going into the history of the influenza epidemic. It encircled the globe, visited the remotest corners, taking toll of the most robust, sparing neither soldier nor civilian, and flaunting its red flag in the face of science."

If anyone might be expected to write about the epidemic it was Vaughan. He was an epidemiologist, one whose professional life centered on understanding the causes and courses of disease, and a medical professional who had been a witness to one of the worst epidemics ever to strike the face of the earth. But instead of dealing with the plague, he seemed almost compelled to quickly give a nod to it and then move on to something easier to talk about. If men like Vaughan did not want to remember the flu, who did?

Military doctors? But they were just as reticent as Vaughan. Several eminent physicians who went to France and wrote memoirs of their experiences failed even to mention the flu. They could not possibly have been unaware of the epidemic, notes University of Texas historian Alfred W. Crosby. American soldiers in France were devastated by the flu. For example, in the fall of 1918, 100,000 American soldiers each month were passing through the Allied Expeditionary Force's center in St.-Aignan-sur-Cher. The flu's toll among them, Crosby writes, was so overwhelming that medical officers had seen nothing like it "since inoculation had controlled that ancient destroyer of armies, typhoid fever." In one division, the 88th, which arrived in Héricourt, France on September 17, 1918, and fought on the front lines from October 26 until the end of the war, the flu killed 444 men. The number who were killed, wounded, missing, or captured in the war was 90.

Military leaders were acutely aware of what the influenza epidemic was doing to their troops' ability to fight. On October 3, 1918, General John Pershing sent out a desperate wire asking for troops and supplies. "Influenza exists in epidemic form among our troops in many localities in France accompanied by many serious cases of pneumonia," he wrote. On October 12, he wired again, his tone more urgent, asking for "one base hospital and 31 evacuation hospitals," saying it was "absolutely imperative" and that along with the hospitals he vitally needed "their nurses and equipment."

Toward the end of the war, German General Erich von Ludendorff, despairing that his country might have to surrender, began fantasizing that a miracle would save Germany. The flu epidemic, he decided, would demolish the French Army. But when his own surgeon general told him that was not going to happen, he refused to believe it, even having his conviction conveyed to Kaiser Wilhelm, whose spirits it lifted. The German surgeon general was correct, of course, since the Germans were at least as devastated by the flu as were the French.

One would think that doctors who witnessed the ravages of the flu in the United States could never forget the hospitals whose every room was filled, or the desperate search for anything—a vaccine, a pill, an elixir—that could stem the disease's course. Who could forget the retired doctors conscripted into service and the pleading patients who wanted someone, anyone, to tend to their illness?

Yet biographers of the great men of American medicine did not elaborate on the epidemic. A 539-page biography of Dr. William Henry Welch was illustrative. Welch was one of the leading medical authorities and a man who was summoned by the U.S. Surgeon General to lead a retinue, including Vaughan, to Fort Devens. But his biographer devotes just three paragraphs of the book to the flu, though it describes the outbreak as one of "the most destructive epidemics of military history." After Camp Devens, Welch continued to tour military camps "with new vigor, trying to fight what he admitted was the most serious situation that had yet arisen." But the book gives no further details and the great epidemic is dismissed as quickly as it is mentioned.

Military historians were similarly unconcerned. Donald Smythe, in an exhaustive biography of General John Pershing, devotes one two-sentence paragraph to the flu. It appears in a section on the battles in France in September and October 1918: "But prospects for the future were bleak, as influenza suddenly broke out, with over 16,000 new cases in the AEF during the first week of October. In America, more than 200,000 were down with the disease, causing General March to suspend nearly all draft calls, quarantine nearly all camps, and considerably reduce training." And that was it. The influenza epidemic never even made it to the book's index.

Equally surprising was the way the 1918 influenza epidemic was discussed in magazines and newspapers of the time. Even the cholera epidemic that started with William Sproat, the epidemic that people attempted to forget, was at least an ever-present subject in the media while it happened. Not so the flu.

Alfred Crosby, puzzling over the epidemic's impact, went to the *Reader's Guide to Periodical Literature* from 1919 to 1921 and counted the column inches devoted to the influenza epidemic as compared with other topics. There were, he wrote, 13 inches citing articles on baseball, 20 inches on Bolshevism, and 47 on Prohibition. Just 8 inches of citations referred to the flu.

Crosby looked at a recent edition of the *Encyclopaedia Britannica*. The 1918 flu got three sentences. He looked at a recent edition of the *Encyclopedia Americana*. One sentence was devoted to the flu, and it said that the epidemic killed 21 million people. "That was a gross understatement," Crosby says. But even so, he remarks, "21 million people? One sentence? Hello?"

When soldiers died of the flu, the cause of their deaths was sometimes hidden in euphemisms, Crosby notes. "At a memorial service for the pandemic dead at Fort Meade, Maryland, the presiding officer read the names of the dead one by one to a massed battalion, and as each name rang out, the sergeant of the man's company saluted and responded, 'Died on the field of honor, sir.'"

When the history textbooks were written, recording for students the events that academic experts deemed important for them to know, once again the flu did not seem to be worth mentioning. Crosby examined college history textbooks, looking for the 1918 flu. He remarked that the epidemic was notable mostly by its absence. "Of the best-selling college texts in United States history, books by such historians as Samuel Eliot Morison, Henry Steele Commager, Richard Hofstadter, Arthur Schlesinger, Jr., C. Vann Woodward, and Carl Degler, only one so much as mentions the pandemic. Thomas A. Bailey in *The American Pageant* gives it one sentence and in that sentence understates the total number of deaths due to it by at least one-half."

Medical scientists are amazed by the great silence, in view of the influenza epidemic's dramatic impact, not just on mortality statistics or an army's ability to fight but in everyday life. They recall that citizens wore white gauze masks in public in a vain attempt to protect themselves. Funerals were limited to fifteen minutes. Coffins were in short supply. Morticians and gravediggers could not keep up with the demand for their services. In Philadelphia, so many bodies had piled up in the morgue that the embalmers said the conditions were "so offensive" that they would not enter it. Public gatherings were prohibited in many cities and some places made it a crime to cough, sneeze, or spit in public. In Washington, D.C., even the Supreme Court adjourned so that, as Justice Oliver Wendell Holmes put it, they could spare lawyers from having to enter "this crowded and infected place." And Washington hospitals were so crowded that they stationed undertakers at their doors to remove each body as soon as death occurred to make room for another patient. "The living came in one door and the dead went out another," one doctor noted. No one could avoid knowing that a deadly epidemic was stalking the land.

But the flu was expunged from newspapers, magazines, textbooks, and society's collective memory.

Crosby calls the 1918 flu "America's forgotten pandemic," noting: "The important and almost incomprehensible fact about the Spanish flu is that it killed millions upon millions of people in a year or less. Nothing else—no infection, no war, no famine—has ever killed so many in as short a period. And yet it has never inspired awe, not in 1918 and not since, not among the citizens of any particular land and not among the citizens of the United States."

In asking why, Crosby proposes a combination of factors that, he said, acting together accounted for the world's collective amnesia. For one, he argues, the epidemic simply was so dreadful and so rolled up in people's minds with the horrors of the war that most people did not want to think about it or write about it once the terrible year of 1918 was over. The flu blended into the general nightmare of World War I, an unprecedented event that introduced trench warfare, submarines, the bloody battles of the Somme and Verdun, and the horrors of chemical warfare.

Moreover, the epidemic had no obvious dramatic effect. It did not kill a world leader. It did not usher in a long period in which death from influenza was a new and constant threat. It did not leave behind legions of crippled and maimed or disfigured survivors who would serve as haunting reminders of the disease.

His latest hypothesis, he said in an interview in August 1998, is that in the fifty years preceding the 1918 flu, the world had been through one of the most profound revolutions ever to change the course of history: the germ theory of disease. "Every eighteen months, a new pathogen was identified, and it went on for years," Crosby noted. Each discovery drove home the message that science was conquering disease. As the drumbeat of infectious agents continued, people "heaved a great sigh of relief. At last infectious disease was not important anymore," Crosby concludes.

Then came the flu epidemic, which made a mockery of the newfound optimism. And when it ended, Crosby posits, perhaps the most comforting reaction was to forget about it, to push it to the back of humanity's collective consciousness as quickly as possible. To "see no evil, hear no evil."

One group of people, however, were haunted by the flu, even if their concerns weren't widely expressed in the popular press. They were the scientists, the medical researchers, who could not rest until they understood what had caused this plague, how it had spread, and how to stop it if it came again. In the years to come, they would recall that flu on two frightening occasions and they would undertake massive public health efforts because, they feared, an influenza virus like the one from 1918 might be coming back.

Medical scientists began their quest to understand what was causing the epidemic literally at the bedsides of the dying in 1918. They continued their efforts for more than a decade until, at last, a man who had been a college student in 1918 stumbled upon a clue that would drive the rest of the century's research.

# 3

# FROM SAILORS TO SWINE

The sixty-two men were in a difficult position. They were sailors from the U.S. Navy Training Station on Deer Island, in Boston Harbor. They ranged in age from a callow youth of fifteen to a seasoned man of thirty-four. Each had been convicted and imprisoned for crimes he had committed while he was in the service. Now, in the month of November 1918, while the influenza epidemic was starting to wane in Boston, a group of Navy officials made these prisoners an offer. Would they agree to be subjects in a medical study that might help scientists understand how the flu was spread? Would they agree to allow doctors to try to give them the potentially deadly disease? If they said yes, they would be pardoned for their crimes.

It seemed a Faustian bargain. Yet the doctors who were hoping to conduct the study, Dr. M. J. Rosenau, who directed the laboratory at the Chelsea Naval Hospital, and Navy Lieutenant Junior Grade J. J. Keegan, reasoned that the sailors could be their best hope of understanding the flu and saving the lives of millions of others.

Today, such studies are illegal. Medical scientists cannot offer inducements like pardons to persuade prisoners to take part in their studies. Although they can award small cash payments to research subjects, they are forbidden from giving anyone so much money or such tempting favors that their compensations might constitute what ethicists term an inappropriate inducement, an irresistible temptation to join the study. Now, more than eighty years after the 1918 flu, people enter studies for several reasons—to get free medical care, to get an experimental drug that, they hope, might cure them of a disease like cancer or AIDS, or to help further scientific knowledge. In theory at least, study participants are supposed to be true volunteers, taking part in research of their own free will.

But in 1918, such ethical arguments were rarely considered. Instead, the justification for a risky study with human beings was that it was better to subject a few to a great danger in order to save the many. Prisoners were thought to be the ideal study subjects. They could offer up their bodies for science and, if they survived, their pardons could be justified because they gave something back to society.

The Navy inmates were perfect for another reason. Thirty-nine of them had never had influenza, as far as anyone knew. So they might be uniquely susceptible to the disease. If the doctors wanted to deliberately transmit the 1918 flu, what better subjects?

Was influenza really so easily transmitted? the doctors asked. Why did some people get it and others not? Why did it kill the young and healthy? Could the wartime disruptions and movements of troops explain the spread of the flu? If it was as contagious as it seemed, how was it being spread? What kind of microorganism was causing the illness?

The normal way to try to answer such questions would be to study the spread of the disease in animals. Give the disease to a few cages of laboratory rats, or perhaps to some white rabbits.

Isolate whatever was causing the illness. Show how it spread and test ways to protect animals—and people—against the disease. But influenza, it seemed, was a uniquely human disease. No animal was known to be susceptible to it. Medical researchers felt they had no choice but to study influenza in people.

Either the Navy doctors were uncommonly persuasive or the enticement of a pardon was overwhelmingly compelling. For whatever reason, the sixty-two men agreed to be subjects in the medical experiment. And so the study began. First the sailors were transferred to a quarantine station on Gallops Island in Boston Harbor. Then the Navy doctors did their best to give the men the flu.

Influenza is a respiratory disease—it is spread from person to person, presumably carried on droplets of mucus sprayed in the air when sick people cough or sneeze, or carried on their hands and spread when the sick touch the healthy. Whatever was causing the flu should be present in mucus taken from the ill. The experiments, then, were straightforward.

The Navy doctors collected mucus from men who were desperately ill with the flu, gathering thick viscous secretions from their noses and throats. They sprayed mucus from flu patients into the noses and throats of some men, and dropped it into other men's eyes. In one attempt, they swabbed mucus from the back of the nose of a man with the flu and then directly swabbed that mucus into the back of a volunteer's nose. In another experiment, the Navy doctors attempted to determine whether the influenza microorganism was a virus or something much bigger, like a bacterium. They forced mucus from sick men through a filter so fine that it trapped microscopic bacteria in its mesh, allowing only the submicroscopic viruses to pass through. They took that filtrate and used it in their attempts to infect the healthy men with the flu.

Of course, the Navy doctors reasoned, just because influenza looks like a respiratory disease does not mean that it cannot be spread by other means. Perhaps, they thought, the infectious microorganism is in the blood. So they drew blood from a man who was ill with the flu and injected the blood directly under the skin of a volunteer.

Trying to simulate what happens naturally when people are exposed to flu victims, the doctors took ten of the volunteers onto the hospital ward where men were dying of the disease. The sick men lay huddled on their narrow beds, burning with fever, drifting in and out of sleep in a delirium. The ten healthy men were given their instructions: each was to walk up to the bed of a sick man and draw near him, lean into his face, breathe in his fetid breath, and chat with him for five minutes. To be sure that the healthy man had had a full exposure to the sick man's disease, the sick man was to exhale deeply while the healthy man drew the sick man's breath directly into his own lungs. Finally, the flu victim coughed five times in the volunteer's face.

Each healthy volunteer repeated these actions with ten different flu patients. Each flu patient had been seriously ill for no more than three days—a period when the virus or whatever it was that was causing the flu should still be around in his mucus, in his nose, in his lungs.

But not a single healthy man got the flu.

It was hard to believe. What was this new plague that swept through military bases like wildfire, killed young men in hours or days, left morgues piled high with bodies, and yet seemed impossible to transmit by all the ordinary ways that diseases are spread?

Perhaps something was wrong with the experiment. Maybe the Boston sailors were somehow inured to the flu—perhaps they had had it, recovered, and built up an immunity. Or maybe those healthy sailors just happened to be naturally immune to influenza. With every disease, there are some people who seem impervious. When the Black Death swept through Europe, killing as much as half of the population in some areas, there were people who never became ill. When cholera devastated Europe, some remained healthy, even though they had eaten the same contaminated food or drunk the same water teeming with cholera bacteria as others who succumbed to the disease and died. There were dedicated doctors and nurses who spent their lives working in leper colonies and never caught the disease. Maybe those Boston sailors were the lucky ones, the people who could never become ill with flu because they had some inborn protection against the disease. But even so, what were the odds that every single one of those sailors would have been immune to the flu?

Another group of doctors decided to try their hand at infecting military volunteers and once more sought men who might agree to take part in their experiments in return for a pardon for crimes they had committed while in the service. This time the studies took place in San Francisco and this time the controls seemed even tighter. There was no way, the doctors reasoned, on the basis of their patients' case histories, that these healthy subjects could have inadvertently been exposed to the flu before the experiments began. Therefore, they could not have developed an immunity to the disease.

The subjects were fifty sailors from the Naval Training Station on the island of Yerba Buena. They had spent a month on the island and not a single man had had the flu. They had been isolated from the epidemic while it raged in the city, and they should have been as free from exposure as a man could be.

The healthy subjects were taken to the Angel Island Quarantine Station in the western part of San Francisco Bay, a few miles beyond the Golden Gate Bridge. And once again, the military doctors tried as hard as they could to give the men the flu. The experiment was the same—inoculate the well with mucus from the sick, inoculate them with blood from sick men, put the healthy in close proximity to the ill. To everyone's surprise, the results were identical. Not one of the volunteers became ill.

Scientists were stunned. If these healthy volunteers did not get infected with influenza despite doctors' best efforts to make them ill, then what was causing this disease? How, exactly, did people get the flu?

The search to find the cause of the flu was fast becoming an international effort, involving hundreds of scientists who were willing to try anything. Some even tried experimenting on themselves. One German investigator asked flu patients to gargle, then filtered the gargled liquid and sprayed it into his own throat and the throats of his assistants. They got some symptoms of flu, although not the full-blown disease. The problem was that they could not prove that they had gotten ill from the throat spray and not because they were living in a community where the flu was raging.

Others tried to transmit the flu to healthy volunteers, conducting experiments much like those in Boston and San Francisco. In Japan, three doctors, T. Yamanouchi, K. Skakami, and S. Iwashima, even seemed to succeed in experiments that took place from December 1918 until March 1919. The question they asked was whether flu was spread by a virus or by bacteria. The healthy subjects included doctors and nurses who volunteered themselves for these grim experiments.

To see if influenza was spread by a virus, the investigators used mucus and blood from flu patients that they had passed through a fine filter to remove bacteria. They dropped filtered sputum from flu patients into the noses and throats of six healthy people. They filtered blood from flu patients and dropped it into the noses and throats of six healthy people, they injected sputum filtrate under the skin of four healthy people, and they injected filtered blood from flu patients under the skin of four healthy people.

The Japanese scientists also tried to transmit influenza with bacteria, dropping a variety of bacteria found in the sputum of flu patients directly into the noses and throats of fourteen healthy people.

The results were exactly as would be expected if flu was spread by a virus: volunteers who had never had the flu and who were inoculated with the filtered blood or mucus got the flu; those infected with the bacteria did not. Those volunteers who had already had a case of the flu did not become ill. But scientists said they were not convinced. Influenza was rampant in Japan. Doctors and nurses would surely have been exposed. The scientists could not be sure that the only way the subjects could have come in contact with the flu was through their experiments. And the reported results were too good, too clean. One hundred percent of those who were susceptible and who were exposed to filtered material, even filtered blood inoculated under their skin, became ill? No one exposed to bacteria from the sputum of flu patients got sick? It seemed too incredible to believe So the quest went on.

Many scientists doggedly attempted to give the flu to animals—monkeys, baboons, rabbits, and guinea pigs. But the research results were conflicted and confused, and no one could claim to have indisputably found the cause of influenza.

In the meantime, the U.S. Public Health Service was trying a different tack to uncover the flu's secrets. Its scientists gathered all the data they could on the epidemic's spread, trying to figure out where it had come from and where it went. They relied on preliminary reports of influenza from September until October 1918, producing a map of the disease's path. The result was striking. Simultaneously, it seemed, the illness cropped up all over the country, much too quickly, the researchers said, to be explained away by travelers or troops taking the flu with them when they moved about.

The health center's report in December 1918 highlights the dilemma: "The outstanding fact, perhaps, is the extreme rapidity with which the epidemic spread after it attained the proportions of an epidemic in the first areas affected. The epidemic became nation-wide in the four or five weeks after it appeared in an epidemic stage in the first localities affected. A fact of scarcely less importance is that the disease reached an epidemic state in a number of localities in the central, northern, southern, and western sections at about the same time as it did in the area along the northeastern coast."

Details of the flu's spread were an abiding puzzle. The disease appeared in nine Army camps in the first week of the pandemic, in states as far apart as Massachusetts, New York, Virginia, South Carolina, and Georgia. In the second week, it appeared in thirteen more camps, including Texas, Kansas, Louisiana, Illinois, Maryland, and Washington.

Moreover, scientists noted, the flu's mortality rates peaked in Boston and Bombay in the same week. But New York, just a few hours from Boston, had its peak three weeks later. And cities much farther from Boston than New York, like Omaha, Memphis, Baltimore, and Montreal, had peaks in their epidemics a week before New York's peak. As one scientist, Richard Edwin Shope, noted years later: "In many respects, the epidemiologist had an easier time getting the pandemic disease transferred from Boston to Chicago, for instance, than he did getting it the remaining thirty-eight miles from Chicago to Joliet. If pandemic influenza spreads from sick to well by contact at the first opportunity as it is assumed to do in explaining the rapidity of its spread over long distances, then it should diffuse with reasonable rapidity over short distances. Yet it does not seem to do so."

Could it be, the Public Health Service report posited, that somehow the organism that caused the epidemic was already present, but unrecognized, in various parts of the country before the deadly disease appeared? Were there smoldering embers of the 1918 flu throughout the country that, for an unknown reason, simultaneously sparked into a raging fire?

. . .

It was not the first time that such a question had arisen. In epidemic after epidemic, doctors and scientists had wondered how the disease could move so quickly through a country, hopscotching over some towns while felling others. There was no epidemic in historical memory like that of 1918, but even epidemics that are just typical influenzas raised troubling questions.

After an influenza pandemic of 1789, a young American doctor named Robert Johnson puzzled over how the infection could spread so far and wide, and so quickly. That epidemic came the year George Washington was inaugurated as President, two decades before the first steamboat crossed the Atlantic, and three decades before the first steam train took its first run. Illness spread so fast that it did not seem possible for it to have been transmitted by person-to-person contact.

"Influenza appeared at London between the 12th and 18th, at Oxford in the third week, and at Edinburgh on the 20th of May," Johnson wrote. Even more puzzling was the spread of the flu on ships at sea, he said. In 1782, a large British fleet set sail for the Dutch coast, with all of the men in perfect health. "Towards the end of May, the disorder first appeared in the *Rippon*, and in 2 days after in the *Princess Amelia*. Other ships of the same fleet were affected with it at different periods: Some indeed not until their return to Portsmouth about the second week in June."

"The present received opinion," Johnson wrote, "is that this species of Catarrh [influenza] arises from contagion, which possibly may be true; yet to my mind it appears no easy matter to conceive how the disease can spread so far and wide in so short a space of time as we perceive it does, or how it can affect persons many miles apart, at the same time, where there had been no previous direct or indirect intercourse—if propagated only by a matter arising from the body of a man laboring under it." Johnson decided that influenza must arise from some sort of changes in the atmosphere but that, once it got started, it could spread from person to person.

Such ideas also persisted after a flu pandemic of 1847. One doctor wrote: "What I wish to point out now is the fact that the Influenza pervades large tracts of country in a manner much too sudden or simultaneous to be consistent with the notion that its prevalence depends exclusively upon any contagious properties that it may possess."

For a brief moment at the turn of the century, however, it seemed that the mystery of the flu had been solved. The flu, many thought, was caused by bacteria.

The man who found what was believed to be the flu microorganism was Dr. Friedrich Johann Pfeiffer, an eminent scientist who was head of the research department at Berlin's Institute for Infectious Diseases. He was credible and he was careful, a man whose word was believed. Pfeiffer made his flu announcement in 1892, just after the last major influenza epidemic before 1918, saying he had isolated a bacterium, which he called *Hemophilus influenzae* and that others dubbed Pfeiffer's bacillus, from the respiratory tracts of people ill with the flu in the epidemic of 1890. Although Pfeiffer failed in his attempts to transmit influenza to animals by infecting them with the microorganism, he convinced most of the world that he had indeed found the cause of influenza. The problem, however, came in the 1918 epidemic.

In the first wave of the 1918 flu epidemic, doctors diligently looked for Pfeiffer's bacillus in the respiratory tracts of patients. To their great surprise, however, they almost never found it.

Then came the second wave, the killer strain of flu. This time, doctors discovered, most—but by no means all—of the flu victims were infected with Pfeiffer's bacteria. But the observations made no sense. If Pfeiffer's bacillus caused influenza, it should have always been present, in all the flu victims.

The findings, most thought, were just too muddled to be convincing. Although some leading scientists remained convinced that Pfeiffer's bacillus caused influenza, more were shaken in their belief in the bacteria. After two decades of assuming that this bacterium caused the flu, it now seemed not to be the cause after all. Now, in the midst of an epidemic that was more deadly than had seemed possible, the enigma of the flu remained. The hundreds of studies conducted during the 1918 epidemic, when the flu microorganism, whatever it was, was at least still present and infecting people, were to little avail. And as the epidemic waned and died out, the flu microorganism presumably died out too, gone from the face of the earth, at least temporarily. The mystery of the 1918 flu was beyond the powers of science and medicine to solve.

"About all that can be said is that the role of the organism was more controversial after the smoke of the 1918 studies had cleared than it had been before," wrote Richard E. Shope, the man who eventually helped solve the problem.

Richard E. Shope was an Iowa farm boy turned doctor who, in retrospect, seemed perfectly prepared by every circumstance of his life to tackle the puzzle of the 1918 flu.

Shope was seventeen years old when the Boston sailors were taking part in the flu transmission experiment. He had grown up in Des Moines, the son of a doctor, and started to work when he was just ten years old milking cows and taking care of poultry and other farm animals. He loved the outdoors, and spent every summer vacation of his childhood and adolescence at Woman Lake, Minnesota, hunting and fishing with his brother. Even as an adult, living on the East Coast, he would return to the lake as often as possible. His personality, too, seemed to fit his Midwestern roots and his time. His tastes were simple, a friend said, and he had "a vivid sense of humor. He was fond of recounting anecdotes, and they became more remarkable and rather less credible with each time of telling."

When it came time for college, Shope went to Ames, Iowa, intending to register in the Iowa State University's School of Forestry. But the office of the registrar at the forestry school was closed when he arrived, so he signed up as a premedical student instead. He entered college in the fall of 1918.

Shope earned his medical degree in 1924, but he had no intention of settling in as a general practitioner in an Iowa farming community. He wanted to do medical research, so he set off for Princeton to study the treatment of tuberculosis at the Rockefeller Institute, a research facility that was, at the time, in the small university town. He loved the rolling countryside, the bucolic little oasis in central New Jersey, a train ride away from New York City—just fifty miles—but light-years away in its atmosphere. One of Princeton's great attractions for Shope was what he called his "gentleman's farm," near the university, where he and his wife, Helen, lived for decades, keeping a cow and chickens and growing vegetables. Shope was one of those sober, yet brilliant, scientists who, a friend said, was possessed of "wide knowledge, common sense, integrity, and complete sanity," traits that stood him in good stead as he ventured into the muddy waters of influenza research.

It began when Shope was working with his mentor, Paul Lewis, at the Rockefeller Institute. A towering figure in the study of infectious disease, Lewis became interested in hog cholera and, knowing Shope's experience with hogs and their diseases, sent Shope back to Iowa to investigate. Hogs there were stricken with cholera; on the vast hog farms in a state where hogs outnumbered people, it seemed a perfect place to understand the disease. But in the autumn of 1928, Shope came across the disease that was to become a consuming passion of his life: swine influenza.

Unlike human flu, swine flu had not been noticed—if, indeed, it even existed—before 1918. But then suddenly, in the autumn of

1918, at the start of the flu pandemic, millions of pigs in the Midwest became ill with a severe respiratory infection and thousands died almost overnight. In fact, animal flus were not unheard of. As long ago as the sixteenth century, there were reports of diseases that sound like flu in horses, for example. The flu that struck the pigs in the Midwest, however, was different. Entire hog farms were decimated.

Whatever the disease was, it looked like influenza. The pigs got runny noses, fevers, even the watery eyes of human sufferers. And the timing of the epidemic's appearance, in the midst of the human influenza outbreak, seemed to some to be more than just a coincidence. It was traced to the Cedar Rapids Swine Show, held in Iowa from September 30 to October 5, 1918, when sick pigs housed with healthy animals soon spread the disease. When the show was over and the newly infected pigs were shipped back to their farms, the disease was seeded throughout the Midwest.

One inspector for the Division of Hog Cholera Control of the U.S. Bureau of Animal Industry, J. S. Koen, said flatly that he was convinced that the swine disease and human influenza were one and the same and that the pigs had gotten the disease from humans. Moreover, he said, there were reports that farm families had also been infected by pigs. Koen named the disease "swine influenza."

Yet in a time when scientists were casting about for an animal that got influenza in order to study the human disease, no one took up the study of influenza in the pig. In fact, many said that Koen was wrong and that pigs were not getting the flu. The animal industry in particular did its best to discredit Koen's hypothesis. Hog producers were afraid that the public might become frightened by the thought that pigs could have human influenza and would refuse to eat pork. But Koen vigorously defended his analysis. He was "a fiery little man," Shope said, and was not about to back down under pressure from the hog industry.

"I have no apologies to offer for my diagnosis of 'flu,'" Koen explained. "Last winter and fall we were confronted with a new condition, if not a new disease. I believe I have as much to support this diagnosis in pigs as the physicians have to support a similar diagnosis in man. The similarity of the epidemic among people and the epizootic among pigs was so close, the reports so frequent, that an outbreak in the family would be followed immediately by an outbreak among the hogs, and vice versa, as to present a most striking coincidence, if not suggesting a close relation between the two conditions. It looked like 'flu,' it presented the identical symptoms of 'flu,' and until proved it was not 'flu,' I shall stand by that diagnosis."

Swine influenza infected pigs annually after that first epidemic in 1918, varying from year to year in its severity but returning regularly with the onset of winter. The swine illness, however, soon faded into the background, of little interest to most scientists studying human flu. But Shope, with his upbringing on Iowa farms and his familiarity with pig diseases, was intrigued. He read Koen's papers and could not dismiss the connection between swine flu and the human pandemic. Could it be, he wondered, that the long-lost 1918 flu organism was still extant, living on in pigs who had acquired the disease from people? If so, Shope and Lewis reasoned, if they focused on the new swine flu epidemic, they had a perfect opportunity to figure out what was causing the swine—and the human—disease.

So the experiments began: Look for microorganisms in mucus secretions taken from sick animals. Try to find microorganisms that were present in sick but not healthy animals. Then try to transmit influenza by giving just those microorganisms to healthy pigs.

At first, it appeared as if it was going to be too simple. In the respiratory tracts of the sick animals was a bacterium that looked exactly like Pfeiffer's bacillus, that controversial, all but discredited bacterium that was once thought to have caused human flu. Perhaps Pfeiffer was right after all.

Shope was struck by the association of the bacteria with the disease. "Frequently, no organism other than this influenza-like bacillus could be recovered from the lungs or bronchial exudate of infected animals," Shope said. "Here then in swine influenza was an organism like that believed by many to be responsible for influenza in man. The problem of determining the etiology of swine influenza seemed simple at this stage, for while the bacillus, which we named *Hemophilus influenzae suis*, was not easy to cultivate, it could always be isolated from cases of the experimental diseases by appropriate methods. In addition, there were numerous cases in which it was the only organism that could be isolated."

The next step was to see if healthy pigs inoculated with the bacterium became ill—an experiment very much like that conducted on the sailors in 1918. If Pfeiffer's bacteria caused swine flu, it should be straightforward to transmit the disease by giving pigs the purified bacteria. So Lewis and Shope tried infecting a pig by dropping the bacteria into its nose. The animal got sick and it seemed to be ill with swine influenza. Moreover, when Lewis and Shope looked for the bacteria in the pig's respiratory tract, they found it. The very first experiment was a resounding success. And the investigators crowed with delight.

"Naturally, we were elated," Shope said. But, he added, "our joy was short-lived because when we repeated the experiment in a second pig we failed to obtain an infection. The animal remained perfectly normal and no lesions suggestive of influenza were to be seen when it was killed after a period of observation. Four other pigs inoculated intranasally with pure cultures of the organism likewise remained normal, and we began to doubt that *H. influenzae suis* caused swine influenza after all." They repeated the experiment—dozens of times—and never again did a pig become ill.

The next year, pigs living on farms in the Midwest again were struck with an influenza epidemic. Once again, that meant that the infectious microorganism was there for the discovering, ready to be found by scientists who were sufficiently clever, or lucky. Lewis and Shope tried again to find the cause of the flu, examining the mucus of sick pigs, searching for the microorganism that caused this disease. But once more all they could isolate was *Hemophilus influenzae suis*. They tried again to transmit the flu to pigs with Pfeiffer's bacillus. Again, they failed.

Shope was intrigued as much as discouraged. The pig flu was turning out to resemble the 1918 flu, at least in the mystery of its transmission. "We were at this stage of the game, in almost the identical predicament regarding the role of *H. influenzae suis* that investigators of human influenza had been regarding the Pfeiffer bacillus at the close of the 1918 pandemic," he remarked. "We had an organism that was regularly present in the disease, which was frequently the only organism to be found in relationship to the respiratory tract lesions, but which, administered in pure culture, failed to produce the disease."

But Shope had one advantage over the doctors in 1918 who were trying to infect humans. In the experiments with military prisoners, doctors could not even infect volunteers with "the crude, supposedly infectious secretions" from the patients' noses and throats. But with the pigs, Shope could transmit the disease with the secretions from sick animals. In fact, he noted, pigs that had not become ill when he gave them pure *Hemophilus influenzae suis* bacteria would develop influenza when he inoculated them with nasal secretions from sick pigs. There was something in that pig mucus that caused the flu. And if there was something there, Shope and Lewis were determined to find it.

Shortly afterward, Paul Lewis became infected with yellow fever while working with the virus in his lab and he died. Bereft at the loss of his mentor and friend, Shope decided nonetheless to continue the flu work on his own. He returned to an old idea. He would abandon the notion that influenza was caused by a bacterium like Pfeiffer's bacillus and ask again whether the disease was caused by a virus and, if so, whether he could isolate it.

The yellow fever virus had been the first human virus, discovered in 1899 by Walter Reed and studied intensively by Lewis and five other Rockefeller University scientists who themselves got infected and died of the disease in the course of their work. The yellow fever virus discovery was, of course, a discovery based on an argument by exclusion. Scientists did not know then that a virus is simply a bundle of genes, in the form of DNA or RNA, wrapped in a coating of proteins and lipids. They were unaware that the protein coat of a virus lets it dock on to cells. They had no idea that the lipids on the virus's surface help it slip inside a cell. They had not yet discovered that genes are made of DNA or RNA. They had not yet invented electron microscopes that would let them see viruses. But they had a fine sieve with holes so small that every known microorganism would get trapped. Viruses, however, were so tiny that they passed through the filter. When viruses were isolated in this way, scientists knew they were present because they could infect animals with the filtered fluid.

Shope's idea was straightforward: He would look for a virus causing swine flu by filtering the mucus and bronchial secretions from sick hogs through a filter that would allow only viruses to get through. But as so often happens in science, an idea that seems easy in theory turns out to be difficult in practice.

Shope had little luck. He painstakingly collected mucus from sick pigs. Then he filtered the material. He dropped the filtrate into the noses of healthy swine. But the pigs did not get the flu. At best, they got a fever or cough, but none developed the combination of symptoms—drippy noses, muscle aches and pains, high fevers—that are the hallmarks of influenza. And there was still that dogged bacterium, *Hemophilus influenzae suis*, that kept turning up in sick pigs. How could Shope ignore the bacteria when they always appeared in the ill animals? How could he assume that a virus was causing the flu when he could not transmit the flu with filtered mucus from sick pigs? The cause of swine flu remained elusive.

"Here, instead of one agent that could be looked upon as of possible etiological importance, were two such agents," Shope said. "The bacterium could not be completely ignored, for, while it was apparently perfectly harmless for swine, its constant presence in so many samples of infectious material from the field and its persistence on serial passage through experimental swine strongly suggested that it must play some role. Neither could the filterable virus be accepted as the cause of the disease because, while it unquestionably possessed pathogenic properties for swine, the mild illness that it caused was certainly not swine influenza. Considered in the light of views current that an infectious disease was caused by a single agent, we had reached a point in our experiments where it appeared that we had one too many under suspicion."

Shope wanted the simple solution—a single infectious agent, whether virus or bacterium. But he was beginning to wonder if there was no such thing as one microorganism that caused the flu. It was a situation that is common in medicine. Diseases that are poorly understood—heart disease, for example—are said to be "multifactorial," meaning that if there is a single cause, no one has found it. Shope was reluctantly joining the multifactorial camp when it came to swine influenza. What would happen, he asked, if he gave the pigs a combination of infectious agents, if he inoculated them simultaneously with the filtrate and the bacteria? He tried it, and, to his surprise, the pigs got not only influenza but a severe pneumonia. The most reasonable interpretation, he decided, was that there was no single cause of swine influenza.

Both the virus and the bacteria needed to be present, acting in synergy to cause the disease.

In England, meanwhile, three scientists, Professor Wilson Smith, Sir Christopher H. Andrewes, and Sir P. P. Laidlaw were working on the mystery of human flu, trying to isolate a virus that caused the human illness. A flu epidemic was in progress at the time—nothing like the 1918 flu in its deadliness, of course, but flu nonetheless.

Theirs was the usual strategy—to look for the infectious agent by taking liquid washed from the throats of people ill with influenza and filtering it in search of the putative influenza virus. These investigators eschewed the strategy of the Navy doctors during the 1918 flu, who tried to infect healthy men with secretions taken from the sick. Instead, they worked with laboratory animals attempting to infect them with human influenza. Others had tried this tack before, of course. But Smith, Andrewes, and Laidlaw were undeterred, and they found an animal that succumbed to the human flu—the ferret.

Ferrets are small and sometimes vicious mammals, relatives of weasels, and not the usual laboratory animal. They had become an offbeat choice of some British scientists who found that they were uniquely susceptible to dog distemper. Unlike dogs, ferrets died when they were infected with canine distemper. The distemper work and the discovery that ferrets were the perfect animal for distemper studies were by-products of flu research. British scientists had reasoned that the symptoms of dog distemper at least superficially resembled those of human influenza and decided that if they could understand distemper perhaps they could understand the flu. They found the distemper virus, but their initial hypothesis proved wrong: the distemper virus turned out not to be related to whatever was causing influenza. Then again, since ferrets were so susceptible to distemper, Smith, Andrewes, and Laidlaw reasoned that it was worth a try to see if ferrets could also get the flu.

The British scientists began modestly, inoculating two ferrets with the filtrate from human influenza patients. Within two days, both of the animals became ill, developing fevers and runny noses and all the outward symptoms of human influenza. It was a promising start, and demanded to be followed up, the investigators decided.

So Smith, Andrewes, and Laidlaw moved their experiments to a research facility run by the British government at Mill Hill, England, where they could do their work with rigid scientific controls. That meant making sure that the ferrets could not possibly become infected with influenza by any means except through the filtrates of mucus and other secretions taken from human patients. The animals lived in complete isolation in a special building. And no scientists or visitors were allowed in, unless they were disinfected, their clothing along with them. The disinfectant was the acrid-smelling household cleaner known as Lysol.

To come into the research building, you had to wear rubber boots and an overcoat, which was hosed with Lysol as you walked by. Then you sloshed through a passageway that was three inches deep in the liquid. By the time you reached the ferrets, you reeked of Lysol but, presumably, were germ-free.

Once they began their studies, the investigators focused on a single strain of influenza, a decision that saved them money by allowing them to exactly replicate their experiments without worrying about variations between influenza viruses. The flu virus they used was isolated from Wilson Smith himself, who had gotten the flu when a sick ferret sneezed in his face. That strain, called "WS" after Smith, still exists.

The ferret experiments were a resounding success. The first question was: Could the investigators transmit flu to ferrets with bacteria-free filtrates from sick people? They could. Then, they asked, could they infect a healthy ferret with influenza by dripping filtrate from the sick ferret in the nose of a healthy animal? Yes. They could even transmit the flu by putting a sick ferret in a cage with a healthy one. What about *Hemophilus influenza*, the infamous Pfeiffer's bacillus? It turned out not to be a factor. Neither it nor other common bacteria transmitted influenza to the animals. Even when the scientists inoculated ferrets with a combination of Pfeiffer's bacillus and the filtrate, the bacteria seemed unimportant. The animals had essentially the same disease as they had when they were inoculated with only the filtrate. So Shope's troubling findings did not seem to hold up. Pfeiffer's bacillus did not seem to be amplifying the symptoms of flu.

Moreover, the British scientists discovered, they could protect ferrets from influenza by first mixing the filtrate with serum from people or ferrets who had already had the flu and then dripping it into a healthy ferret's nose. The immunity that builds up after an influenza infection was in the blood and could block the flu virus.

To complete the picture, the British scientists found a connection between human flu and swine flu. The ferrets were susceptible to both diseases. Smith, Andrewes, and Laidlaw could give ferrets influenza by inoculating them with filtrates from pigs that had swine influenza.

Shope was intrigued and his interest was further piqued after he met Andrewes for the first time. Shortly after he did his flu work with the ferrets, Christopher Andrewes visited Shope in Princeton and the two scientists compared notes. This, Andrewes said, was the beginning of a long and close friendship between the two men.

Shope decided to dig deeper into the mystery of swine flu, using ferrets and asking again some of the questions that had nagged him. He confirmed that ferrets did indeed develop flu when he inoculated them with the swine flu filtrate. But these were not easy experiments to do. The snarly animals with sharp teeth did not want to sit still while Shope dripped flu-containing liquid into their noses. Shope decided to anesthetize the ferrets before inoculating them with swine flu. To his surprise, he discovered that when he did the experiments this way, the flu virus caused not just influenza but also a severe pneumonia, one that caused the lungs to swell with watery blood and that sometimes even killed the animals. It looked for all the world like a typical case of the 1918 flu.

Finally, Shope wondered whether it made any difference if *Hemophilus influenzae suis* was given along with the filtrate. It did not. Whatever the *Hemophilus* bacteria were doing, they were not, apparently, causing influenza.

But was it just another fluke, a trick of nature that the British influenza and the swine flu could be transmitted to ferrets that year? Were all influenzas caused by something that could pass through a fine filter that left even bacteria behind? Dr. Thomas Francis, a scientist at the Rockefeller Institute who went on to run the field trials of the polio vaccine in the 1950s, found an answer.

It was 1934 and flu was raging in Puerto Rico. Francis isolated the organism that caused it using Shope's technique, so he could infect ferrets with the virus. He also found that Shope's discovery of what happened when he gave swine flu to the anesthetized ferrets held true for the human flu virus: if Francis passed the virus from one ferret to the next, anesthetizing them when he infected them, the flu virus gave the animals a serious pneumonia along with the flu.

In a final coup, the British scientists Smith, Laidlaw, and Andrewes and, independently, the American scientist Francis found that they could give white mice the flu if they took the flu viruses from sick ferrets. The virus, apparently, changed somewhat while it was growing in ferrets so that now mice were susceptible to it. And mice were not just susceptible—they developed the worst sort of flu. It was accompanied by an often fatal pneumonia.

Shope asked the next obvious question: Were the human flu virus and the swine flu virus one and the same? With the discovery that mice and ferrets could get flu from both human and swine viruses, the pieces were in place to get an answer.

Shope gave ferrets and mice influenza by inoculating them with a human flu virus. When the animals recovered, he tried giving them swine flu. They were immune to it. Then he tried the opposite experiment, giving ferrets and mice swine flu. After they recovered, they were immune to human flu.

It was a test of the antibodies that are made following a flu infection. These complex proteins pour out of white blood cells to serve as an armed force against the virus. When a flu virus antibody comes across a flu virus, it attaches itself to the virus and blocks it so that the virus cannot infect cells. Shope had found that swine flu antibodies could protect ferrets and mice against human influenza and vice versa.

Shope discovered that the human and swine viruses were not identical. He tried laboratory experiments in which he mixed swine virus with antibody-containing serum from pigs that had recovered from swine flu. The antibodies in the serum completely inactivated the virus, so that if he inoculated ferrets or mice with it, the animals never became ill. Antibodies in the serum of pigs were only partially effective in inactivating the human flu virus.

Shope tried the experiments again, but this time he started with serum from people who had recovered from influenza. The serum completely inactivated the human influenza virus, he discovered, but only partially inactivated the swine flu virus. At least the flu strains that were around a decade after the 1918 epidemic were slightly different from swine flu.

There was a burning question left unanswered, however. What about the 1918 flu? Was it more than a coincidence that the first known swine flu epidemic emerged in the midst of the worst human influenza pandemic in recorded history? Could it be that humans gave pigs influenza in 1918 and that it remained in the animals, perking along at a low level ever since?

The virus that caused the human 1918 flu pandemic was gone, of course, vanishing with the pandemic. No one had saved it—they had no idea how to save a virus and, in fact, medical scientists of that time did not even know that it was a virus that had caused the 1918 flu. But, of course, diseases do leave traces behind, in the antibodies that remain in serum, ready to block that virus should it return.

Shope and the British scientists Smith, Andrewes, and Laidlaw had an idea. They would look at the antibodies that remained in the serum of people who had survived the 1918 flu. Were those antibodies an exact match for the swine flu virus? Was the 1918 flu virus the same as the swine flu?

They persuaded volunteers of all ages in both countries to roll up their sleeves and give blood. Then the scientists took the precious straw-colored serum from that blood to their labs and asked what kind of antibodies it contained and whether, in particular, the flu antibodies of people who had been alive in 1918 were different from the flu antibodies of people who had been born after that terrible pandemic.

The results were unequivocal. Both in London and in the United States, people who had survived the 1918 flu had antibodies that completely blocked Shope's swine flu virus. People who were born after 1918 did not have those antibodies.

Even Shope was surprised. "The results were quite unexpected," he said. The studies, he explained, "indicated that almost all adults had undergone an infection with a virus of the swine influenza type." The most likely explanation, Shope said, was that he had found the footprints of the 1918 flu. They were the antibodies left behind in the flu's survivors and they showed that the human flu virus was, in fact, one and the same as the swine flu. The 1918 influenza virus, it seemed, may have survived after all, in the bodies of pigs.

Not everyone agreed. Francis, Andrewes, and others suggested another hypothesis. It might be, they said, that the antibodies that block swine flu resulted not from a previous infection with that virus but as a generalized response to repeated infections with a variety of influenza viruses. The investigators proposed that as people grew older and had been infected time and again with various human flu viruses, they developed a sort of universal flu antibody, one that effectively latched on to and inactivated a variety of influenza strains, including the swine flu.

Shope stuck to his hypothesis, arguing that the alternate one "seems awkward to me, and it also seems particularly fortuitous, too much so, that this general influenza antibody, so called, should happen to react selectively with a virus of the antigenic composition of swine influenza virus." And, he noted, 35 out of 112 people he tested had antibodies that blocked the swine flu virus but not human flu viruses. How could that have happened? "In these, then, the added assumption would have to be made that the 'specific' human virus antibody disappeared while the 'non-specific' swine virus antibody persisted: a highly artificial rigmarole to go through to avoid a simple and direct explanation," he insisted.

As the 1918 flu receded further and further into the past, continuing studies of antibodies confirmed Shope's conviction. Only people who had been alive in 1918 and living in areas where they had been exposed to the flu that year had antibodies that could block the swine flu virus. Thomas Francis, for example, discovered in 1952 that he could not find any antibodies to swine flu virus in people born after 1924 and that the antibodies were most prevalent among people born between 1915 and 1918. Another study looked at people living in Alaska, where the 1918 flu had decimated some villages and spared others entirely. People who had lived in villages struck by the 1918 flu had antibodies that blocked the swine flu virus; those who had lived in villages that escaped the flu did not.

The mysteries of the 1918 flu were beginning to be solved. There was a swine flu connection—probably people had given the flu to pigs, scientists proposed, and the virus may have remained in pigs, lying dormant until, one day, it might strike back at people. But there remained the question of when, and whether, the deadly flu virus might come back to infect the human race. And if it did, how could we get advance warning?

Other pressing questions about the flu's virulence and its spread were still unresolved. No one knew, for example, why it was so deadly. Perhaps nature would give scientists a clue. The 1918 epidemic came in two waves, a mild flu in the spring of 1918 followed by the killer flu in the fall. And it seemed that the two flu strains were closely related. Infection with the first strain protected against the second

This was noticed early on by the military in particular. The 1919 *Annual Report of the Surgeon General of the U.S. Navy* stated that "many men of the Navy who had influenza in the spring or summer of 1918, while in European waters, escaped during the later epidemics, both in Europe and the United States." The same observation was made by the British Grand Fleet, involving British servicemen: "The conclusion is that mild attacks earlier in the year, as a rule, conferred immunity against the more fatal type of disease which prevailed subsequently."

Colonel Victor C. Vaughan noticed that men who had gotten the flu in the spring of 1918 seemed protected from the second wave of flu in the fall. He pointed out that the 2nd Infantry Regiment was struck by the first wave of the flu in June 1918 while the soldiers were in Hawaii. They were transferred to Camp Dodge in the beginning of August. In September and October, the second wave of the 1918 flu swept through Camp Dodge, sickening a third of the men and killing 6.8 percent. But virtually none of the men who had been in Hawaii in June at the time of the first wave of the flu became ill in the Camp Dodge epidemic.

Vaughan also cited the experience at Camp Shelby, where a division of 26,000 men encountered the first wave in April of that year, with some 2,000 men falling ill. "This was the only division that remained in this country without change of station from April until the fall of 1918," Vaughan remarked. "During the summer this camp received 20,000 recruits. In October 1918 the virulent form of influenza struck this camp. It confined itself almost exclusively to the recruits of the summer and scarcely touched the men who had lived through the epidemic of April. Not only the 2,000 who had had the disease in April but the 24,000 who apparently were not affected escaped the fall epidemic. It appears from this that the mild influenza of April gave a marked degree of immunity against the virulent form in October."

It began to seem that the flu virus had gone somewhere between the first and second waves of the 1918 flu where it mutated into a killer strain. Perhaps, some thought, it had gone into animals.

Shope had another idea, one that remained highly controversial. He suggested that swine lung worms were intermediate hosts for the influenza virus, a place where the virus could hide between pandemics, and that the deadly second wave of the 1918 flu involved the same virus as the one that caused the first wave, or one closely related. In essence, he thought that the virus had reemerged virtually unchanged in the fall of 1918. That is why people infected in the first wave were protected from the second wave. But, he said, the difference between the first and second wave of the flu was not the virus but a hanger-on, the old nemesis of flu researchers, Pfeiffer's bacillus. Shope was convinced that people were dying of influenza in the second wave of the 1918 flu because they were infected with both the virus and the bacterium, which amplified the virus's effects.

These theories today are largely discarded, remnants of another time. *Hemophilus influenzae* is now known to cause a bacterial meningitis in children, and there is a vaccine to protect them against it. But Shope left behind a legacy that was something of a double-edged sword. On the one hand, his work helped fling open the doors to influenza research. He found the swine flu virus and made the connection between it and the human flu of 1918. He promoted the hypothesis that the 1918 flu virus lived on in pigs.

At the same time, Shope's brilliant work, through no fault of its own, was to spark one of the greatest product liability fiascoes in U.S history. In 1976, because of his theories about swine influenza, the United States government ended up trying to immunize all Americans against swine flu. The decision was made by President Gerald Ford on advice from the nation's most eminent scientists. A young soldier had died of a swine flu, igniting a fear that a deadly flu like the one from 1918 might be spreading to humans.

It turned out that there was no swine flu epidemic—although no one could have known that at the time. Meanwhile, thousands of people became convinced that the swine flu vaccine caused medical problems ranging from paralysis to fatigue to nearly any chronic disease, initiating hundreds of thousands of legal claims. By the time it was called to a halt, the swine flu immunization campaign had generated a legacy of mistrust of flu vaccines and a mistrust of scientists, who were harshly criticized—unfairly, they say—for crying wolf. It became an object lesson for scientists on the dangers of taking the 1918 flu too seriously, Crosby says. After the swine flu disaster, the 1918 flu "became something to avoid," if you were a virologist, Crosby added.

Another mystery of the 1918 epidemic remains a stubborn and troubling footnote to this medical history. Why were those sailors in 1918 immune to the flu? Perhaps many had gotten mild cases of flu in the first wave of the pandemic and so were protected during the second wave. Perhaps the patients dying of the flu who were supposed to infect the healthy men had already passed the stage in their disease when they were infectious. A few days after a flu virus takes hold, the virus is mostly gone, wiped out by antibodies, the immune system's defense. The symptoms of flu continue in absence of the actual virus, caused by the immune system's attempts to protect the body by flooding the lungs with white blood cells and fluids.

There is yet another reason why those early experiments to transmit the flu might have failed, notes Dr. Edwin Kilbourne, an eminent flu researcher now working at New York Medical College in Valhalla. Perhaps, Kilbourne says, those healthy men had had the flu, but neither they nor anyone else knew it. In his studies, Kilbourne found that at least 7 percent of people who were infected with influenza had a symptomless flu. The virus lives in the lungs and multiplies. The infected people develop antibodies that fight off the flu by blocking the virus. But for reasons that are still completely unclear, these people never develop symptoms.

In the aftermath of the 1918 influenza pandemic, scientists realized that despite the unexpected triumphs of their research, despite their discovery of the footprints of the 1918 virus in swine flu and their continuing investigations, they still lacked the direct knowledge they needed to really understand that terrible virus and to protect people from it if it came again. One thing was required, and it seemed almost impossible to obtain. Scientists had to get hold of the actual 1918 influenza virus itself.

# 4

# A SWEDISH ADVENTURER

When Johan V. Hultin looks at his happy life, his financial success as a pathologist, the elegant and airy apartment that he shares with his wife high on San Francisco's Nob Hill, when he surveys his collection of pre-Columbian art and recalls his vacation home in the nearby Sierra Nevadas and his travels to virtually every country in the world, he knows exactly the moment that his fate was set in motion. It started off as an ordinary day in January 1950, when Hultin, a twenty-five-year-old visiting postgraduate student, set off as usual to do his research at the University of Iowa. But by that afternoon, Hultin was on his way to becoming the Leif Erikson of the 1918 flu.

Hultin was intrigued by the flu, of course, but he had never thought that the pandemic of 1918 would define his life. Instead, he had come to Iowa from his native Sweden, where he was studying medicine at the University of Uppsala, as part of a grand adventure to see the world and to take advantage of a special program in Swedish medical schools that allowed students to leave halfway through their studies to pursue other interests and to return with no loss in standing. Hultin's plan was to study the body's immune reaction to ordinary influenza. If anyone had asked him about the 1918 flu, he would have parroted the conventional wisdom of the time: That horrible strain of the virus was long gone and had left no traces that would enable microbiologists to figure out what made it so deadly.

In the years since Richard Shope had made the connection between swine flu and the influenza virus of 1918, scientists had made steady progress in understanding influenza viruses, but had not come close to figuring out what made one virus a killer and another a dud.

In 1936, they learned that they could grow flu viruses in fertilized hen's eggs, a discovery that made influenza one of the most popular viruses for scientists to study. It was so easy—you did not need a ferret or even a mouse. All you needed was an incubator and patience. The idea was to inject the virus into the amniotic fluid surrounding the chick embryo. When the embryo breathes, it draws the amniotic fluid in and out of its lungs. The virus grows in the embryo's lungs and then is secreted back into the amniotic fluid when the embryo exhales. If a virus grows in an egg, the normally clear amniotic fluid turns cloudy with viruses within two days.

Soon afterward, scientists found that there are different strains of the flu and that there are even broad classes of flu strains. Most human flu viruses are of strains they dubbed A, and these A viruses mutated rapidly, so someone who has recovered from a bout with flu can get it again the next year because the virus will have changed enough to evade the immune system's defenses. Another type of flu, called the B strains, also could infect humans but seemed less likely to mutate from year to year. Why? No one knew.

In 1941, scientists learned that flu viruses have a distinctive protein, which they called hemagglutinin, because it makes red blood cells (which contain the oxygen-carrying molecule hemoglobin) agglutinate, or clump. If they mixed serum containing flu viruses with red blood cells, the viruses would attach themselves to the red cells, hooking them together in a lattice that would sink to the bottom of a test tube, forming a red button that was a sure sign that the virus was present.

In 1944, Americans were the first in history to be immunized against influenza, with flu shots made of viruses that had been grown in eggs, then killed so that they could not cause an infection. That meant that if scientists knew in advance that a new flu epidemic was brewing, they might be able to stop it with a vaccine. In 1947, the newly formed World Health Organization created a worldwide surveillance system to provide an early warning of flu epidemics.

For Hultin, however, these developments were interesting, but not compelling. When he arrived in Iowa, he was more interested in adventure. Why Iowa? Hultin said that his microbiology professor in Sweden suggested the medical school in Iowa City and Hultin was intrigued because, he said, "it was in the middle of the U.S., where Swedish immigrants had settled. It had a good medical school and its microbiology department was outstanding." It was also, of course, the university made famous by Richard Shope.

On that fateful day in 1950, Hultin woke up early, as was his habit, in the small room in Iowa City that he and his wife, Gunvor, were renting. After breakfast, he set off immediately to work on his influenza project in a lab at the university. The lab was a large room filled with graduate students, each at his or her bench, each conducting experiments that would form the basis for a postgraduate degree. Every now and then, eminent microbiologists would pass through town and would be ushered into the lab to see the busy students at work. That morning, as Hultin looked up from his lab bench, he noticed that the head of the microbiology department, Roger Porter, was escorting William

Hale, a well-known virologist from Brookhaven National Laboratory. Porter would stop at each student's bench and tell Hale what the student was doing. When he came to a student whose work was particularly interesting, Porter would pause and explain the work more fully.

When the men came to Hultin's lab bench. Hultin recalls, Porter said simply, "Here's Johan Hultin. He's from Sweden and he's working on the influenza virus." Then they moved on.

A few minutes later, however, Porter brought Hale back to Hultin's bench, remarking, "Bill, you've got to see what this fellow Hultin here has made." It was a jury-rigged device involving a Bunsen burner, one of those ubiquitous gas burners used in labs to heat liquids, and an alarm clock that Hultin had modified to solve a vexing problem. In those days, every student in the large room had a Bunsen burner and every one had a mechanical alarm clock perched on three legs. They would set their alarms to tell them when it was time to turn off their Bunsen burners, and the alarms were going off constantly. The students not only had to put up with the constant shrieking of the alarm bells; they often did not realize that the alarm going off was their own, causing them to ruin experiments.

Hultin had spoiled quite a few of his own experiments before he came up with a way to solve the problem. He realized that he could use a piece of brass tubing to connect the rotating arm on the back of the alarm clock, which cocked the alarm's bell when the time was up, to the gas valve that controlled the flow of gas through a rubber tube to the Bunsen burner. When the clock's alarm sounded, it would automatically shut off the gas to the Bunsen burner and extinguish the flame.

"It really worked," Hultin said. His first test of the device was a resounding success. "I set the alarm to five minutes, turned the Bunsen burner on, and then I walked away. In five minutes, the alarm went off and the gas valve shut. It was so simple." When the other students saw what he had done, several asked him to make the device for them, too, which, of course, Hultin did.

Porter thought that Hale might be amused by Hultin's device. "He asked me to demonstrate," Hultin said, explaining to Hale that it would not take long. "Then he asked me to set the alarm for ten seconds. After ten seconds, the alarm rang and the gas turned off. Hale just stood there. He said, 'My God. For eighty years people have been ruining experiments around the world. And no one thought of this simple solution.'"

The men walked away, with Hale shaking his head in amazement. Two hours later, a secretary walked up to Hultin in the lab and told him that Porter had invited him to have lunch with Hale and several other select students and faculty members. It was a lunch in the faculty dining room, the sort of gathering that universities often arrange as an opportunity for the most promising or most senior graduate students to meet leading scientists from elsewhere. Such visitors might prove to be valuable contacts for students in the future. In the meantime, faculty members can have an opportunity to exchange their latest thoughts and data with intellectual leaders in their field. Hultin's faculty advisor was present, as was Porter, four other faculty members, and three graduate students. And, of course, Hultin, who was invited in recognition of his invention.

That day, Hultin said, the conversation around the table at lunch ranged widely, focusing on science but flitting from topic to topic. Then Hale made an offhand remark about the 1918 influenza epidemic. It was a remark that was to change Hultin's life.

"Everything has been done to elucidate the cause of that epidemic. But we just don't know what caused that flu. The only thing that remains is for someone to go to the northern part of the world and find bodies in the permafrost that are well preserved and that just might contain the influenza virus."

Hale was saying that if someone could find bodies of flu victims that had been frozen since the day they died, the intact virus that had killed them might be chilled to a state of suspended animation. If those corpses had remained buried in permanently frozen ground in the northern regions, the influenza virus that was in their lungs might still be alive. And if the virus could be brought back to the laboratory and revived, scientists might study it and figure out why it was so deadly. They also might be able to produce a vaccine against the disease.

The suggestion went unremarked. "It was a very short comment, it took just ten or fifteen seconds," Hultin said. "Then he went on to something else." But Hultin was transfixed. Of all the people in the room, of all the people in the world, he was uniquely positioned to do just what Hale had suggested. By chance alone, Hultin knew where to find permafrost, he knew how to find tiny outposts where people lived in regions where the ground was frozen year-round, he knew how to get permission from relatives to exhume flu victims buried in permafrost, he knew how to take tissue samples and how to preserve them, he knew how to coax viruses to grow in the laboratory, and he was working with a professor who was a renowned leader in influenza and could help him figure out the virus's secrets.

"I knew this was for me," Hultin recalled.

It was a strange odyssey that led Hultin to his position that day, as the one person who could fulfill Hale's vision. Everything in his past and all of his passionate interests, developed in reaction to his life in Sweden, had prepared him.

He was born in Stockholm and grew up in a wealthy home in the suburbs of the capital city. His father owned an import business. Hultin had two sisters, but one died when she was six months old from an infection in her finger that spread deadly bacteria through her bloodstream. The other sister died in an accident when she was thirty-two.

Despite his family's wealth, Hultin describes his early childhood as disadvantaged. His parents were extremely conscious of social class and forbade Hultin from playing with many children who lived in the village because their social standing was lower than his. Hultin remembers chafing at the restrictions of his society and determining to rebel.

When Hultin was ten years old, his parents divorced and his mother remarried, this time to Carl Naeslund, a prominent professor of medicine at the Karolinska Institute in Stockholm, the place that selects the recipients of Nobel Prizes. For many years, Hultin's stepfather was chairman of the committee that chose the winner of the Nobel Prizes in medicine.

"He was a good man to me. He understood how I thought," Hultin said. Living with Naeslund and his son "was a very happy period in my life," Hultin adds. To this day, Hultin keeps a picture of Naeslund in his office.

Hultin was impressed by Naeslund's scientific accomplishments—Naeslund even had a bacterium, *Actinomyces naeslundi*, named after him. He also was struck by his stepfather's other abilities. Naeslund spent much of his spare time building a house, and an extraordinary house it was. It had a Mediterranean feeling to it, with an indoor garden featuring palm trees, fig trees, and even a pond. After Naeslund died, the royal family of Sweden bought his house and one of the royal princes lived there. Eventually, the royal family sold the house to a Norwegian shipping magnate, who resides in it to this day.

Naeslund also built a summer house, a log cabin on an island in the Baltic Sea. And his eager acolyte, Hultin, followed along, learning to build and work with wood and soaking up a love of microbiology along the way

Naeslund understood Hultin's frustration with Sweden's rigid class society and his burning desire to work, to do manual labor with the common people, and to see what life was like for those who were not wealthy. When Hultin was sixteen, Naeslund helped him to get a summer job as a lathe operator at a factory that made autoclaves—sterilizing machines—for hospitals. Hultin was overjoyed—the job would allow him to be different from his schoolmates and he would incur his mother's disapproval, which he relished. Also, the work would be a challenge, and even then Hultin loved challenges.

That summer, Hultin would come home each evening, his clothes spattered with machine oil, which he saw as a badge of honor. "I did not want to change my clothes. I wanted to show everyone that I was a worker," Hultin said. The neighbors objected to the sight of him. He remembers one woman in particular, the widow of the Swedish ambassador to Great Britain, who called his mother, outraged, and afraid that Hultin had turned into a laborer.

When Hultin was nineteen, he graduated from high school (Swedish high schools went on for two years beyond American schools), was admitted to the University of Uppsala to study medicine, and got a summer job as a longshoreman. He loved the idea of the job—physical labor, exotic cargoes, a job that once again was utterly at odds with his social class—and tried to pass as a commoner. He said nothing about where he came from, taking pride in his strength and ability to work long hours at an arduous job. "I could work as hard as they," Hultin said, recalling that he had been athlete of the year in his high school, running sprints and 400-meter relays, throwing the discus, and also competing in the high jump.

But the other longshoremen instantly knew Hultin was an outsider. It was his accent, he said. "The Swedish I spoke was equivalent to English spoken at Oxford or Cambridge, England," he explains. And the other workers despised him, accusing Hultin of slumming and taking a job from someone who really needed it.

One day, Hultin said, a group of four longshoremen lifted a heavy crate above their heads, positioning Hultin in the middle.

One gave a signal and suddenly, the four men walked away, leaving Hultin alone, poised with the heavy box over his head. Terrified, he said, "I jumped away and the thing crashed to the ground, opening up." How, he wondered, could he continue working there? Help came from an unexpected source.

"The foreman called me over and said, 'I understand you have problems.' I said, 'Yes.' He said, 'I figure you can add and subtract, with your good high school education.' Then he showed me the record keeping he was doing. He asked me to help him and he said, 'If you do this for me, I will keep those guys off your back.'" Hultin, of course, agreed, and although the other workers were never friendly, they stopped harassing him, because the foreman had made it clear that Hultin was protected. Later, Hultin found out the rest of the story. "It turned out that the foreman was a chronic alcoholic who showed up for work with a small suitcase full of beer. He could not keep the records," and was in danger of losing his job, Hultin said. That was why he had approached Hultin.

In 1946, as soon as World War II ended, Hultin left his summer job and set out to see the world, hitchhiking through war-torn Europe. He ended up in North Africa in 1948.

Travel "was a great adventure," Hultin remembered, but it also had its terrifying moments. The worst came when he arrived in Cairo, two days after Israeli planes had flown over the city and dropped bombs. The city was in chaos and lawlessness reigned. The Egyptian military had requisitioned all means of transport, including trains and buses, and there was virtually no lodging available. Hultin went from hotel to hotel, from rooming house to rooming house, trying desperately to find a place that would take him. Finally, he found a room. It was his, he discovered, because the man who had been staying there had gone out and never returned. "He was murdered, one of an average of a hundred murders a day, I was told," Hultin said. "Anyone who looked like a foreigner could be killed. The owner of the hotel kept telling me, 'Don't leave. Don't leave.' And in this room was the man's suitcase, reminding me of what might happen."

Hultin stayed holed up in the hotel room for a week, finally insisting on leaving to go to the market nearby. The proprietor gave him a small Koran to carry for safety, which Hultin has kept ever since. But when he went to the market, an Egyptian man accosted him, accusing Hultin of being a British spy. "I said in my poor English, 'No. no. I'm a Swedish medical student.' He felt cold rivulets of sweat trickle down his body. His heart pounded and he remembered that suitcase in his hotel room, the luggage of the previous tenant who went out and never came back.

"In a panic, I spoke German," Hultin recalled. "That was my first foreign language and I spoke it rather fluently." In German, he insisted again, "No, I'm Swedish." Then one man in the group of Egyptians surrounding Hultin said that no, Hultin must be a German soldier who had escaped from a prisoner-of-war camp.

"I said again, 'No, no. I'm a Swedish medical student.' Then finally I got my head on right and agreed with them that I was a German soldier." The hotel owner arrived then and pulled Hultin away. Hultin escaped from Cairo and made his way back to Sweden, working in the engine room of a Swedish freighter.

Back home, Hultin settled down to medical school. Halfway through his studies, he married his high school sweetheart, Gunvor. The two had begun dating when they were sixteen. She was of Hultin's social class—her parents were Norwegian but lived in Sweden because her father owned the National Cash Register Company in Sweden. Gunvor was studying radiation biology and had learned the new method of radioisotope tracing from its inventor at the University of Stockholm.

Not long after their marriage, Hultin proposed going off to Iowa for six months, to see the United States. Gunvor was happy to go along. She even got a job at the university, which was overjoyed to have her as one of the few experts in the isotope method.

The Hultins left in the spring of 1949, planning to see the country before school started in the fall. They took a boat to America, arriving at Ellis Island after ten days at sea. Once in Manhattan, they met a former classmate who invited the Hultins to stay with him for a couple of days and drove them around the city. The Hultins were the original naive visitors. When their friend showed them a sign that said "coin laundry," Hultin recalls, "I never asked him what it was—I knew it. Americans are so worried about germs that they have their coins cleaned."

After New York, Hultin and his wife flew on a DC-3 to Tucson, Arizona, where an aunt and uncle of Gunvor lived. They visited for a week, marveling at the majestic deserts with the towering cactuses, brilliant blue sky, and florid sunsets. Then it was time to see the rest of the United States. They borrowed a car from Gunvor's aunt and uncle—a small 1947 Studebaker—and borrowed some money, planning to spend as little as they could and to camp out rather than stay in hotels. "We wanted to see all the states," Hultin said. They were in a hurry to gulp in the vast land that is North America before school started in September, believing that they were only going to stay in the United States for six months. "I had our return tickets to Sweden in my pocket," Hultin explained.

And so they set off, ending up by visiting—or at least physically setting foot in—each of the forty-eight states and all but two of the provinces of Canada. Then they drove north, following the romantic vision of traveling until the road ended. It ended in Alaska, which was unexplored wilderness, with few roads and few people. The end of the road was Dawson's Creek, the site where the Alcan Highway began. The Hultins, of course, wanted to go on. By a stroke of fantastic luck, they could. Just two days before, the Canadian government had opened the highway to civilians—it had previously been restricted to people with military permits. It was not a highway like the paved and smooth interstate highways of today. Instead, it was a lonely and hazardous road, and to travel on it, Hultin and his wife had to show that they had a spare fuel pump, a fan belt, and inner tubes. Theirs was just the tenth civilian car to journey along the highway.

Hultin.and his wife drove for days, seeing almost no one. Parts of the road were dirt, and rutted with huge potholes, deep and wide. "We stopped at a truck stop and the truckers would look at the car with amazement. They had never seen a car with such small wheels," Hultin said. The land was pristine. "You can't imagine how many fish. The streams were full of trout. Anytime you looked into a stream you would see ten, fifteen, maybe twenty graylings. Every time I threw out my fly line I caught one," Hultin recalls with wonder. "You could imagine what America looked like when only the Indians lived there," he said.

Every evening, and some lunch hours too, Hultin would set off with his fly rod. At the same time, Gunvor would start a fire in their cookstove, knowing that in the fifteen minutes it took to get the stove going, Hultin would be back with fresh fish.

The only problem was the huge mosquitoes with their high-pitched whines and their unerring ability to home in on human flesh, truly malevolent insects that plagued the Hultins in their tent at night. They had no choice but to camp out—hotel rooms were so scarce that they were all taken by three in the afternoon in the few small outposts and towns along the highway. They were so expensive that the Hultins could not afford them anyway. But it stayed light until ten at night and neither Hultin nor his wife wanted to stop driving at three.

Finally, they arrived at Fairbanks, a frontier town that met all of Hultin's expectations, and more. "I had an idea about the Western pioneer towns of the gold-mining era," Hultin said. "Fairbanks looked just like I had imagined a Wild West town. It had honky-tonk bars and dirt roads." And it had many hotels. It might, Hultin thought, have accommodations for them. They might, finally, spend a night in a real hotel room.

The problem, however, was that rooms in Fairbanks cost so much. They went from hotel to hotel, asking the price of a room and turning away, dejected, envisioning another night camping out with the mosquitoes. Finally, a hotel clerk told them to go to the University of Alaska in Fairbanks, a couple of miles outside the city. Classes were over for the summer and the students had gone home, so the reservations clerk thought the Hultins might be able to rent a room in a dormitory.

The clerk was right. The Hultins got a small room with two beds in a primitive wooden dorm for married students. It cost them fifty cents a night.

Although the university was closed for the summer, some faculty members remained on campus, working on their research projects. The Hultins soon met one of them, a Norwegian, who struck up an immediate bond with Gunvor and introduced the Hultins to a German paleontologist at the university named Otto Geist. Geist's graduate student assistant was on vacation and he needed someone to go along to help him on his digs in Alaska. In return for working for Geist, the Hultins could have their dorm room for free. It was not even a question. A chance to see the Alaskan wilderness? With a paleontologist? The Hultins did not hesitate to say yes.

As they got to know more about Geist, they could hardly believe their luck. He was famous in his field, having characterized a new species of bison, called Superbison, that used to inhabit the Alaskan tundra. He also had a flamboyant streak, lining his fifty-yard-long driveway with the three-foot-wide bison skulls.

"They were museum quality," Hultin said, still wondering at Geist's choice of landscape ornamentation. "When we left after many weeks, he gave us a skull. I still have it."

But that summer, Geist was not out for bison. Instead, he was looking for bones of the earliest horses. He traveled up and down the coast of Alaska's Seward Peninsula, walking or taking a dogsled or flying and landing on beaches even where it seemed impossible for a plane to land. He was so well known and well liked that when Eskimos in isolated villages heard the noise of an airplane's engine, they would clear the beach of driftwood so that Geist could touch down. A gregarious man going forth alone, he had made close friends, it seemed, in every remote village.

Hultin and his wife spent weeks with Geist, drinking in his stories, learning paleontology, and discovering the violent and beautiful Alaskan wilderness. They dug up mammoth tusks, one as long as sixteen feet. Hultin found a huge jawbone of a mammoth, but reluctantly turned it over to the university while longing to put it in his car and take it home.

When the summer was over and the Hultins drove back to Iowa. Gunvor began working in the radiology department. Hultin began his microbiology studies, taking up his daily position at his lab bench, setting up his Bunsen burner so that the alarm clock would turn off the gas. Hale came to visit; Hultin was invited to lunch. And then Hale made his fateful remark about the 1918 flu.

As soon as the words were out of Hale's mouth, Hultin began dreaming of doing exactly what the virologist had suggested—finding flu victims who had been buried in permafrost and extracting the 1918 virus from their tissue. He knew how to find permafrost in Alaska—the federal government would have maps telling where it was located. He also had a plan for using those permafrost maps to locate Eskimo villages that had three other necessary features: they were built on permanently frozen ground, they had good records of the dead from the 1918 flu, and their residents might allow him to dig up the flu victims' graves. The key was Geist. Geist knew where the villages were and could provide introductions to missionaries who would have historical records of flu victims. Geist could introduce Hultin to Alaskan villagers who might be persuaded to give permission to have their ancestors exhumed in the name of science.

Hultin decided to test the waters at his university. He went to his faculty advisor, Albert McKee, a virologist who was an associate professor of microbiology at the university and who had been at the luncheon with Hale. Without revealing his whole scheme, Hultin simply asked McKee if he remembered what Hale had said. He did and remarked that he thought that Hale had an interesting idea. Hultin oh so casually asked McKee: What if he took on that project? McKee, never one to say no to an idea, Hultin recalled, replied, "Oh yeah. Why don't you think about it."

A couple of days later, Hultin went in to see McKee again, this time laying out his plan. "I told him that I know somebody who knows everyone there along the Seward Peninsula and that I'd done a lot of work with him," Hultin said. McKee, intrigued, encouraged Hultin to get started and see whether he could make the project work.

The first step was to write to Geist. Hultin explained Hale's idea of looking for frozen bodies of flu victims buried in permafrost and asked Geist how to go about finding the record of deaths from the influenza pandemic of 1918. Hultin suspected that the missionaries had records, but he needed help finding them.

Geist wrote back immediately, offering assistance and telling Hultin that he would get names and addresses. Hultin could say that Geist referred him when he wrote his letters.

It took all winter. Hultin corresponded from home, in the evening. Gradually, replies drifted in. Some missionaries said there were no records. When the calamity hit in 1918, it wiped out as many as 90 percent of Eskimos in some villages; nor did it spare missionaries. Those who survived had little time for record keeping. They had to find a way, first, to bury the dead and care for the orphaned children. Mortality statistics fell by the wayside.

Between writing to missionaries and waiting for their replies, Hultin searched for information on permafrost in Alaska. This time, he relied on the brother-in-law of his department chairman, Roger Porter, who was a congressman and who directed Hultin to Army records, giving him temperature readings in the ground and in the air every month of the year for years on end. "I could draw the permafrost line on a map," Hultin recalled.

Finally, putting all the data together, Hultin could make his plans. "I knew where the missions with good record keeping were located relative to the permafrost line. Then I found that only three places would work. They were the only places where there were adequate records and where the burial had been in permafrost," Hultin said.

By then, a year had passed, and it was time to seek funds to pay for an expedition to Alaska. In March, Hultin applied to the National Institutes of Health for a research grant. "A month went by and I didn't hear anything," Hultin recalled. Then another month elapsed, and another. Roger Porter wrote a letter to the institutes asking the reason for the delay. He was told that there were many scientists seeking funds and that Hultin's application would be examined and ruled on in good time.

Finally, Porter called his brother-in-law, the congressman, asking him to find out the real reason for the delay. He discovered that the Army, with the help of the National Institutes of Health, had seized upon Hultin's idea and was planning its own expedition to Alaska to do exactly what Hultin had proposed to do. Except the Army mission, whose code name was Project George, was going to be classified and, Hultin later learned, was to cost $300,000. But discovering that his idea had been appropriated only made Hultin more determined to go ahead with his plans and to beat the Army at its own game.

"When we found out, Roger Porter went to the university medical school's central scientific fund and requested ten thousand dollars," for the Alaska mission, Hultin said. "He got it immediately and we were on our way within a couple of days. We knew that the military had resources to get there quickly and we were going to get there ahead of them." The team consisted of Hultin, McKee, and Jack Layton, a pathologist at the University of Iowa. They would meet Geist at the University of Alaska in Fairbanks. Together, the group hoped to find a tissue sample from a 1918 flu victim and to bring the virus back alive.

The Iowa group flew to Fairbanks via San Francisco and Seattle, carrying with them wide-mouthed thermal jugs that they had filled with dry ice. They intended to put their samples in the jugs and keep them frozen during the journey home.

It was early June when they arrived, and they planned to stay at the dorm at the University of Alaska for the usual fifty cents a night. They decided that Hultin would be the scout, going to the three sites where it was possible that flu victims were still preserved in frozen ground. He would telegraph the others if he found well-preserved bodies. He would travel to the remote villages with a bush pilot.

But when the men arrived in Fairbanks, the sky turned charcoal gray and it began to rain. Their hearts sank. It would be impossible for a bush pilot to land on a beach in such weather. All they could do was wait for the rain to end and hope it ended soon.

The rain continued, pounding the city and much of Alaska. Day after day the four men hung about the city. Every morning they awoke to the drumming sound of rain. The dirt roads were mud now, the tundra a swamp. The rain kept on falling.

One morning, the men realized they had a new problem. They had come prepared to take the frozen tissue from the corpses and to carry it back to Iowa, planning to pack it in dry ice that they had brought in the sterile thermos bottles. But as they waited for the rain to end, their dry ice was beginning to evaporate.

"The longer we waited, the lighter the thermal jugs became," Hultin said. "In no time—it was just a week or two—there was no dry ice left. And try to find dry ice in Alaska. It's tough."

They searched for a supplier, but to no avail. They tried to think up another way to bring back the samples still frozen, but could not find one.

"That was a calamity, right there," Hultin said. Finally, he stumbled on an idea. "I was sitting there, really depressed. Then I remembered that when you use carbon dioxide fire extinguishers, that white cloud that comes out, that's dry ice, in powder form." He nearly whooped with elation. He had found the perfect solution. His friends cheered and the men rushed off to the closest fire department, asking where they could buy fire extinguishers. They were sent to a store where they discovered they could buy as many fire extinguishers as they wanted, as many, they decided, as they could carry through the tundra. So they bought a half dozen. All except one was small, but the final one was enormous, weighing about thirty pounds. The problem was solved.

At last, two weeks after the gray weather had descended upon Fairbanks, the sun shone again and Hultin set off to see whether any frozen corpses of victims from the 1918 flu remained.

Hultin described it as "one of the great adventures of my life."

Hultin landed in Nome, a port city and the first place in Alaska to be hit by the 1918 flu. There was permafrost, there was a big Lutheran mission with good records of deaths, and there was a cemetery where the flu victims had been buried in the fall of 1918. On paper, at least, "everything seemed right," Hultin said.

But when he saw the cemetery he was bitterly disappointed. "The description of the gravesite that I had read was not what I was looking at," he recalled. A river passed through Nome, and in the thirty-four years since the 1918 flu, it had changed its course so that it came very close to the mass grave. An edge of the cemetery was now adjacent to the river and there was almost no chance that the ground would still be frozen. "I dug a test hole near the grave. There was no permafrost," Hultin said. The only thing to do was to move on.

Hultin later learned that the Army's classified mission arrived at that same graveyard in Nome just ten days later. They landed in Air Force transport planes with diesel generators to run freezers and they set up camp. Then they started to dig. But it did not take them long to discover that all that remained of the 1918 flu victims was their skeletons. With the permafrost gone, the bodies had decayed and there was no soft tissue—and no virus—left.

The last survivor of that Army mission is Maurice Hilleman, an august virologist who later became director of the Merck Institute in West Point, Pennsylvania, which is a research center for the pharmaceutical giant Merck & Company. Hilleman was working for the Army at the time, at the Walter Reed Institute in Washington, where his job, he said, was to study the 1918 flu with the hope of preventing the next pandemic. He went along on that trip to Nome, and he remembers just what Hultin discovered. "The bodies were in such an advanced state of deterioration that no live virus was found." Looking back, he said, it was no surprise. "What is permafrost? It's supposed to be permanent frost. But what happens is that this year there may be a thaw down for two feet. The next year the ground may thaw down for six or eight feet." Over the years, he said, "you would expect at least one warm cycle. And why should they bury those bodies deep enough so they would even reside in permafrost? That was the big problem we had."

The Army expedition was restricted to Nome because it needed to fly in its planes with diesel generators to power its freezers. Hultin had no such limitations since he planned to keep the tissues frozen in his thermos jars that were packed with dry ice. So while the Army group was digging in Nome, Hultin had set off for his next site—a small town called Wales, which was the westernmost point on the continent, just across the Bering Strait from Siberia.

Taking off from Nome under an overcast sky that looked more and more threatening, the bush pilot followed the beach, flying just a few hundred feet above the ground. As the sky turned to thick fog, he had to fly lower and lower. About forty minutes after leaving Nome, the pilot began looking for a navigational marker—a cabin on a bluff.

"Again and again as we circled, he missed seeing it and was therefore forced to fly closer to the top of the bluff," Hultin recalls. "Yes, you guessed it. We were flying right into it, and at the last moment the pilot managed to avoid the end of both of us. We cleared that little hut by fifteen to twenty feet. That shook the pilot and he turned back toward Nome in the dense fog, flying low over the beach."

The pilot tried again the next day—the fog had cleared and Hultin was enjoying the magnificent scenery. This time, as they approached Nome, the single engine of his plane started to sputter and finally died.

"A disturbing silence to say the least," Hultin said, "but the bush pilot reassured me that we were going to land and fix the wire that had come loose from a magneto terminal. Apparently it had happened before, so he knew how to fix it. At that point I should have been extremely concerned, but having heard so many bush pilot stories, I had developed such a degree of admiration for them that I did not have the slightest worry, and, with apologies, I inquired about a place to land. There simply was none—only the dark hostile crags as far as the eye could see. 'There is a frozen lake here somewhere,' the pilot said. And sure enough, even I could see the little white dot about two thousand feet below. As he banked, the engine started. 'Great. The wire fell back on the terminal,' was his only comment, but as he corrected and assumed his course, it again sputtered and stopped. Needless to say, the next twenty minutes or so were very long before he finally set down on the beach at Wales."

Wales had been decimated by the 1918 flu, which appeared in the village after the virus had shown its face in Nome. There had been a variety of stories about how the disease arrived there. One said that a little boy had died while visiting friends in another village. His father went to pick up his body and brought his dead son back to Wales by dogsled in November 1918. The boy had had the flu. And that, some said, marked the entry of the virus into Wales.

Another account claimed that a mail carrier was bringing letters from the port city of Nome to Wales by dogsled. He got sick along the way and died. His dogs, in the meantime, became ravenous and began barking furiously. Hunters heard the dogs, found the mail carrier, and brought his body to Wales, spreading the flu to the tiny village.

The stories never seemed right to Hultin. A body, dead for days, would be unlikely to spread an airborne disease. Finally, in 1998, he heard what he thinks is the real story. He was in an Alaskan village, Brevig, speaking about the flu, when a woman in the audience approached him, telling him that she had grown up in Wales and that her great-grandfather was the mail carrier of the legendary story. He had been transporting mail on his dogsled from Nome, she said, and had fallen ill with influenza on the way. He managed to get to Wales, where he died the next day. A week later, 178 out of the 396 residents of Wales were dead, victims of the flu.

All that was known when Hultin arrived in Wales in the summer of 1951 was that the flu had struck the village with a vengeance. He saw a large cross marking the grave where the bodies of the 178 villagers lay. They had been buried at least six feet down, in permafrost. But the grave, once inland, was now on a bluff overlooking a beach. The beach had shifted in the intervening years. Looking at the warm sun shining on the bluff, Hultin realized that the permafrost was no longer there. Just to be sure he was right, he dug down into the bluff. The ground was soft. It was, he said, "another failure."

Now he was down to his last site, Brevig. It would be his final chance to find frozen victims of the 1918 flu.

But leaving Wales was not easy, because a storm was blowing in from the Bering Strait, making it impossible for the bush pilot to take off from the soft sloping sand of the beach. Hultin and the pilot remained there day after day, growing increasingly impatient. Hultin, ever the adventurer, spent his time talking to the Eskimos, noting with interest that some had lived in sod houses with whale ribs to support the walls.

Days after the bad weather began, the pilot decided to take a chance and to try to fly out of the village. The pilot marked a runway on the beach, creating a circuitous course that avoided places where the sand was soft. Then they got in the plane. The pilot turned to Hultin and explained what would happen. "We are going to pick up speed and then turn right into the wind. We have to have enough speed so that when we turn and lose some speed we can still get into the wind and lift. If it's really bad, we won't make it but we will only end up in the water. The Eskimos will pull us out."

The pilot took off, quickly ran out of runway, turned into the wind, and began his ascent. The plane was so low that its wheels hit the tops of the waves a few times. "I could feel it, bump, bump, bump," Hultin said. But they made it and set off for Brevig.

The beach was soft at Brevig and there was no place for the plane to land. Hultin and the pilot had to land instead at the larger town of Teller, six miles away, which had a beach with firmer sand. Eskimos met them there and took Hultin across the inlet to Teller in a whaleboat made of walrus skin. It was about fifteen feet long and six feet wide, with an outboard motor, and it carried seven people. "I went to Brevig and introduced myself to the missionary," a man named Otis Lee, Hultin continued. "He didn't know I was coming—there was no communication available—so I just showed up. He was a very friendly man and I stayed with him and his wife in the old mission." It was a wooden house built on permafrost but the house had gradually melted the frozen ground, making it buckle. The house was sagging; its floors were not level.

Hultin could hardly wait to visit the gravesite in Brevig, and when he saw it—about thirty feet long and marked at either end by two large wooden crosses, one nine feet high and the other five feel high—he knew that, at last, he had found a place where the permafrost might have remained. There was a good chance that here, in Brevig, he might find still frozen bodies of victims of the 1918 flu.

Seventy-two out of the eighty people living in Brevig had died of the flu in November 1918, so many that even burying them was problematic. Two months passed while the frozen flu victims remained aboveground, a silent testimony to the virus's terrible force. Even if there had been able-bodied people to dig a mass grave, it would not have been easy. Permafrost is nearly impossible to penetrate. The Alaskan territorial authorities hired gold miners from Nome who had the equipment to dig a hole in the frozen ground; by contract, the bodies had to be buried at least six feet deep.

The miners came to Brevig in January 1919, equipped with a steam generator that heated water and pumped steam into a rubber hose attached to a pipe. They hammered the pipes into the ground, thawing it, and enabling them to dig the grave. Then the seventy-two bodies were placed in the grave, the grave was closed, and it was marked with the crosses.

The missionary's wooden two-story house, with six rooms, was converted to an orphanage for the children of Brevig and surrounding villages. Soon it housed about a hundred children whose parents had died suddenly of the 1918 flu.

• • •

But for Hultin, in 1951, Brevig looked like it might be the site he was looking for. Greatly encouraged, he asked Otis Lee for help.

"I told Lee that I needed permission to do the work," Hultin said. "So he called the village council together and told them why I was there and how important it was. There were three survivors there from 1918," three of the eight people in the village who were not killed by the flu. "I asked them through an interpreter to tell the story, as they remember it, the story of November 1918. They did and I said, 'It is now possible to prevent this from happening again. But I need your help. If you allow me to dig in the grave, I will do my best to find some specimens. And once we have the virus, we can produce the vaccine. And the next time the disease comes, you will be immunized and you will not die.'"

The villagers told Hultin he had their permission to dig. The next day, he began, with his pick and shovel.

"I started right in the middle of the grave. I came into permafrost after about three feet."

Digging that frozen ground was very difficult, Hultin discovered. He was alone, and the earth was hard and rubbery. He struck it with his pickax but it did not budge. The Eskimos looked on but did not offer help. Hultin thought about his problem. It seemed to him that the only way to penetrate the earth was to warm it up and melt it a little. And the best way to do that would be with a fire. He brought in driftwood from the beach and built a fire, piling up some twigs and branches and lighting them. As the fire burned, the ground softened, which allowed Hultin to dig and scrape off about two inches of melted ground. He built another fire and scraped off another two inches.

Hultin devised a systematic way to go about his work. He would build a fire on one side of the hole he was digging and warm up the earth. Then he would dig a couple of inches while another fire warmed up the ground on the other side of the grave. He went back and forth, digging first one side of the hole, then the other.

The hole began to grow, inch by inch, until it was clearly becoming more of a pit than a depression in the earth. Then other problems arose.

"As I progressed deeper and deeper, I had to create an air intake and an exhaust site," Hultin said. He had to get oxygen to the fires and he had to make sure the smoke was carried out of the pit. But that was only the beginning of his difficulties. The fires, it turned out, did not burn well because the sides of the pit started to melt from the heat of the fires below and as they melted they would drip into the flames, extinguishing them. Also, the deeper he dug, the scarcer combustible air became, which made it increasingly difficult to keep the flames going.

Hultin, however, was indefatigable. The fires kept going out. He was working in muck and mud. The smoke burned his eyes and choked his lungs. But he kept at it, working sixteen to eighteen hours a day, taking full advantage of the northern summer, in which daylight lasted virtually around the clock. After four days, his hole was about six feet long, three feet wide, and six feet deep. Suddenly he came upon a flu victim.

The first thing he saw was her head, with black braided hair and bright red ribbons. "She was a little girl, about six to ten years old," Hultin said. She was wearing a dove-gray dress, the one she had died in. "I dug a little more to see what the body looked like but I didn't do more than that." It was time, Hultin realized, to call in his colleagues in Fairbanks and get help with this project.

About two days later, the team arrived. They landed at Teller, six miles away, and they carried the small fire extinguishers and the large one with them. Hultin met them there, and as the junior member of the group, he offered to transport the large fire extinguisher across the tundra to Brevig. It was not an easy task. It had rained recently and the ground was spongy and soggy. The tundra was covered with large mounds of grass, making walking difficult under the best of circumstances. And the fire extinguisher was cumbersome. Hultin rigged up a wooden, cradlelike device—like a backpack—to hold the fire extinguisher as he slowly slogged along, sinking about six inches with every step. "I was walking and sinking, walking and sinking, mile after mile with this heavy fire extinguisher. It was very tough going," he said.

But Otis Lee, the missionary, set out to help, journeying across the tundra, riding a tractor to which he attached a flat-bottomed aluminum boat. He met the group halfway and pulled the men and their fire extinguisher on the boat for the remaining three miles to the village.

The four men stayed in the village's one-room schoolhouse, sleeping on the floor on air mattresses. The next morning, they woke up, went to the gravesite, and started to dig. The weather was ideal, sunny and warm, heating the ground enough so the scientists did not have to build fires. They used their pickaxes and cleared a large area of the grave, about twenty-five feet long and seven feet deep. They found not only the body of the little girl but also four more bodies. They stopped there. "We had enough," Hultin said. It was time to remove tissue from the corpses' lungs. If there was a virus to be found, they should be able to find it in that tissue.

The scientists were on the verge of cutting out snippets of tissue from frozen bodies that might be infected with the most dangerous virus ever known. They faced a real possibility of unleashing a new pandemic upon the world. They had no idea whether the virus would be alive, but the whole point of their expedition was to try to revive it.

There were no national or international commissions ruling on the safety of what these men were doing. There were no ethics committees or lawyers deciding how to ethically or legally protect the Eskimos of Brevig or the rest of the world from what might have been a catastrophe. Instead, the scientists were on their own. And their only concern was the science. Rather than worrying that the corpses might infect them, they worried that they might contaminate the corpses with their own viruses and bacteria. But Hultin believed "the risk was minimal," since they were very careful when they removed the small pieces of frozen lung tissue and placed them in the sterile containers. "The risk would be much greater in the laboratory later," he added.

When they dug into the part of the grave where bodies were lying, the four men wore gloves and surgical masks to cover their noses and mouths and they sterilized their equipment to keep from sullying the tissue samples they were removing. They asked the Eskimos to stay away while they removed tissue from the corpses. But that was the extent of their protective measures. They went about their task in a single-minded way, exhilarated by the thought that they were actually doing the hypothetical experiment that Hale had suggested. Their priority was getting the samples back to the Iowa lab where they could be studied.

"In 1951, I was a graduate student," Hultin explained. "I just didn't have enough knowledge of how things spread. That's why I felt comfortable having a professor of virology with us, because he would protect us. We took precautions that were standard at the time, but we were not afraid of getting infected. I can't remember that we even hesitated doing the postmortem." McKee, Hultin added, "had probably spent twenty years dealing with the influenza virus and other organisms. He was an experienced virologist and he did not express fear."

Even though the protective measures that the scientists used in 1951 look primitive now, "at that time it was the best effort." But, looking back, Hultin can't help but shiver. "We should have been more concerned about not starting this pandemic again."

It took the men about two and a half days to dig up the four bodies. When they were ready to remove the lung specimens, they cut out the ribs with rib cutters, instruments that look like pruning shears. Then they removed the chest plate, revealing the lungs. If the virus was to be found at all, it should be present in the lungs. "We probably had a two-inch cube from each lung," Hultin said. "The reason we didn't do more was that we had a limited number of specimen containers," eight-ounce screw-cap jars that had been sterilized. As they got each specimen, the men would put it in a thermal jar and squirt dry ice in the jar from the fire extinguisher to keep it frozen.

Finally, their job done, they closed the grave and telegraphed the bush pilot to meet them at Teller and return them to Nome, where they could catch a flight for Anchorage and then begin the long flight home. The next day, Lee got out his tractor and his flat-bottomed aluminum boat and pulled the men to the landing site.

Hultin was elated. Everything about the trip to Anchorage only fueled his excitement. "I remember the ride back from Nome to Anchorage. We flew close to Mount McKinley. It was sunset and the cloud layers were absolutely flat, with the sun slightly above. The peak of Mount McKinley stood up out of the clouds with an alpine glow. It was a magnificent sight."

The three men from the University of Iowa flew home on a small plane, a DC-3. "We had the containers in the passenger compartment—no one knew what it was. It looked like camping gear," Hultin said. "That was fifty years ago and you can imagine how many times a DC-3 airplane had to stop to refuel. Every time it stopped, we would go behind the building with our fire extinguisher and put some more dry ice in the containers." The fire extinguishers, he explained, "made a terrific noise." The men discharged the extinguishers as far as possible from the airplane. "I didn't want too many questions coming my way."

• • •

In Iowa, it was Hultin's job to try to get the virus out of the frozen lung tissue, using standard methods of virology that are still followed by scientists trying to grow live influenza viruses. He began by grinding up the tissue, suspending it in a salt solution, and spinning it in a centrifuge to separate the virus from debris. He added an antibiotic to the fluid to kill any bacteria present—viruses are impervious to antibiotics—and then was ready to begin the tedious work of injecting the fluid into fertilized chicken eggs. Hultin would very gently cut a one-half-inch square of shell off the egg, revealing the delicate membrane underneath. Into that window he would inject the fluid that, he hoped, contained the 1918 virus, poking his needle through the membrane so that it got into the white of the egg itself. Working with a technician, Sally Whitney, Hultin injected hundreds of eggs. "We spent a month and a half injecting eggs before we used up all of the specimens," Hultin said.

Hultin was restless with excitement, waiting to see if the virus would grow. "I remember the sleepless nights. I couldn't wait for morning to come, to charge into the lab and look at the eggs," he said.

But when he eagerly arrived each morning, the results were always the same. The antibiotics kept bacteria from growing. But the amniotic fluid remained clear. No virus grew.

Disappointed, Hultin nonetheless hoped that his other attempts to resurrect the 1918 flu virus might succeed.

He injected suspensions of lung tissue into the nostrils of guinea pigs, white mice, and, following Shope's lead, into the nostrils of ferrets. The ferrets were difficult to work with. A male technician wearing heavy leather gloves had to hold the ferocious animals while Hultin anesthetized them with ether, swabbing the drug over their noses and mouths to put them to sleep so they could be inoculated with lung tissue from the flu victims. "They were always fighting and they were very powerful little animals,"

Hultin said. But the ferret experiments were crucial. "The literature showed that ferrets were very susceptible to the influenza virus," Hultin noted.

After all his efforts with chick embryos and rodents, Hultin failed. "I had used up all the specimens and I got nothing," he said. "Nothing worked. The virus was dead."

Hultin thought about taking precautions, just in case the flu virus was alive, as he hoped it would be. He and Whitney wore masks and sterile gowns; they worked under negative-pressure hoods, like those over kitchen stoves, where the air is swept up and under the hood into an exhaust duct rather than into the room. Those were the same precautions that McKee had established for working with highly dangerous bacteria that cause tularemia, Hultin noted. They were state-of-the-art at the time.

Those conditions, however, would be considered primitive today. When scientists work with deadly viruses, like the Ebola virus, they work in elaborate, specially designed labs. Before they can enter, they remove all of their clothing and everything that touches their skin, including rings and contact lenses. They change into sterile scrub suits. Then they walk into a room that is under negative air pressure—air cannot flow out of the room, only in. There they are bathed in blue ultraviolet light, which kills viruses.

After that, they tug on latex gloves and tape the gloves to their sleeves and their socks to their pants, sealing them. Finally, they put on pressurized space suits and attach the suits to air hoses. It is a system so elaborate that only a few labs in the world are set up for this kind of research. If this is what scientists think is needed to study the Ebola virus, which, though terrifyingly deadly, is not spread except through direct contact with a victim's blood or body fluids, how strange to think of studying the 1918 flu virus in an open lab under a hood.

But such containment measures had not been invented in 1951, and Hultin did not dwell on the possible danger of unleashing the virus as he did his work. Instead, he obsessively applied every trick of virology to try to get the flu virus out of the lung tissues of the Alaskan corpses.

Hultin never wrote up his results, never published a scientific paper on his failed attempt to resurrect the 1918 flu virus. "I thought, 'I have it all, I can always write it up later,'" he said. By then, his six-month stay had been extended to two years and he was expected to finish his master's degree—imminently. "I was busy with my master's degree project and it was a negative finding," Hultin said. "If it had been positive it would be tremendous, but it was negative."

Hultin expected that that was the end for him in Iowa. He would get his master's degree and go back to Sweden. He had no great desire to return to his homeland, where "in academic life you advance only because someone ahead of you dies or retires." He dreaded returning to the rigid class system and what he describes as "an extraordinarily punitive taxation system and a life full of limiting boundaries" that prevented people from exploring new arenas. He had fallen in love with America, the antithesis of Sweden, he thought. And so, as his two years in Iowa drew to an end, he was a bit melancholy. It had been a great adventure, he wished he could stay, but his time was up.

To Hultin's great surprise, Roger Porter, the head of the microbiology department, asked him if he wanted to continue his studies there as a medical student. Hultin was so certain that he would have to return to Sweden that he had already bought his return ticket. "It took me two seconds to recover from the surprise and one second to make a decision, one second to say yes," he said. "Then I called my wife. I knew she would agree."

"The opportunity to continue medical school was marvelous—I couldn't imagine that something that good could happen," Hultin exulted. There had even been an unwritten ban on foreign medical students at the university, he alleged, or so rumor said. "They had had a foreign student thirty years earlier and had had a problem. I never found out what it was," Hultin recalled. Anyway, with the flood of GIs returning to school and the abundance of bright students from the United States, the university felt no need to admit foreign students. He could hardly believe an exception had been made for him.

Hultin believes he got the offer to continue at the University of Iowa because he had helped the university at a time when it needed a public relations effort. The day he had returned from Alaska, the university had become embroiled in a scandal. A young woman who was a student there and who was the daughter of a prominent alumnus had died and her boyfriend had been charged with murder. The university had tried to keep the incident quiet but it had leaked out. The charges eventually were dropped.

"The university was desperate for good news," Hultin recalled, "and so they grasped for straws with this trip of ours to Alaska. They ballooned it up." The university's public relations office asked Hultin if he would be interested in driving around Iowa giving lectures about the influenza virus and the expedition and showing slides of the site. Of course he complied and had a marvelous time, lecturing at Rotary Clubs and polishing his English along the way.

When school started in the fall, Hultin was struck by the contrast with the University of Uppsala, a place where the medical school was "an easygoing undertaking without the propellant known as fear of failure," he recalled.

At Uppsala, the dean set the tone in his address to the medical students on the first day of class. He said. "Well, let's have some fun in the next couple of years. Don't worry, you will all graduate." The medical school program was supposed to take six years.

But the dean told the new students, "I know some of you will take seven or eight years, but that's not important. I know some people who took ten years. I know one who took fifteen years—they all sort of fell asleep. I know about that. I was one of them."

In Iowa, the students were deadly serious and determined. School to them was a ticket to success, the first entree into a meritocracy where the very best could thrive. Most of the medical students, just returned from fighting in World War II, were mature men, earnest and extremely competitive and astonishingly good students.

The university made it clear that only the best of those students would make it. Other students told Hultin—who entered the school as a sophomore—that on the first day of medical school in Iowa, the dean gave a talk to the freshmen. But it was not reassuring like the speech he had heard in Sweden. Instead, the dean had said, "Take a look at the person sitting to your left and to your right. Chances are that person will not be there four years from now."

Every Friday, at four in the afternoon, the class standings were published outside the dean's door, a list of the students ranked according to their cumulative test scores as of that week. And every Friday, the students would walk with trepidation up to the dean's office to look for their names on that list. Hultin worked as hard as he could. He studied so much that he could not imagine studying more. But nearly every week he failed to rise to the top 20 percent of the class. What was worse was that everyone knew that if their performance was deemed unacceptable they could be asked to leave the school at any time, even at the end of their medical school education. It happened in Hultin's class, at the very end of the fourth and final year of medical school. In the last exam that year, three students failed. One was allowed to take the exam again, and he passed. He got his medical degree. The others were asked to leave, without a degree.

"There were so many who didn't make it," Hultin ruefully recalled. "I worked so hard, but I was never number one—I was not even close. I was number 16 one week in a class of 104 and I thought, 'My, my, my. Imagine these students. Fifteen of them beat me. How can anybody know so much?'"

He did graduate, and eventually ended up in private practice as a pathologist in California, working in San Francisco and Los Gatos for thirty years. He had a good life. In his immaculate home office, with papers and books carefully filed and tucked away in every square inch of space, Hultin has a large map of the world, pierced with pins. The pins represent visits to a region and the map indicates that he has been to every corner of the earth. In his spare time, for the past quarter century, he has been building his vacation home in the Sierra Nevadas. It is a replica of a fourteenth-century Norwegian log cabin. He spends much of his time there now and it is where he keeps the superbison skull that Otto Geist gave him in 1949.

Every now and then Hultin thought back on that memorable trip to Alaska which had sparked his lifelong interest in influenza. "Every time I saw an article about influenza I would read it and file it," Hultin remarked.

Discovery after intriguing discovery was made. Scientists learned how the influenza virus invades cells. Like all other viruses, those causing influenza cannot live by themselves but must enter the body cells, where their sole function is to replicate, commandeering the cell's own molecules and using them as machinery that does nothing but make tens of thousands of viruses. Flu viruses depend upon two sorts of proteins to enter and exit from cells: one, the hemagglutinin that coincidentally makes red cells clump is used by the virus to hoist itself into a cell, and the other, known as neuraminidase, is used by newly made viruses to burst the cell open so they can escape in a spray and infect new cells. The hemagglutinin and neuraminidase proteins protrude from the flu virus's surface, providing targets for the body's immune system, which tries to stop the viral invasion.

It is those two viral proteins, the hemagglutinin and the neuraminidase, that define a flu strain, and scientists began naming strains by their hemagglutinin and neuraminidase proteins. A strain that swept the world in 1946, for example, was H1N1. The next time the flu virus underwent a major genetic change, creating a pandemic, was in 1956, with strain H2N2. The pandemic that arrived in 1968 involved a virus whose hemagglutinin had changed from the 1956 virus but whose neuraminidase had not. It was named H3N2.

In the battle between invading viruses and the defending immune system, white blood cells generate antibodies that lock on to a flu virus's hemagglutinin and neuraminidase proteins, blocking them and so defusing the virus. But it can take days for the body to develop enough antibodies to stop a flu infection, unless, that is, the flu strain is one that has invaded the body before. In that case, the immune system can quickly marshal its forces and block the virus before an infection can take hold. When the flu virus makes dramatic changes in its hemagglutinin or neuraminidase genes, flu victims can be at its mercy, which is why pandemics occur.

But the body also has another defense against the flu, as scientists discovered in 1957, a sort of natural antibiotic that kills viruses, the influenza and others. It is a protein named interferon, secreted by white blood cells, which wrests control of the cells from the viruses, forcing the cells to make a variety of proteins to thwart the viruses. The most important among the proteins is PKR, for phosphokinase RNA, which prevents viruses that use RNA as their genetic material—like the flu virus—from replicating.

Hultin kept track of all these discoveries and took special note of the influenza pandemics and how vulnerable people were to these viruses.

There had been the pandemic of 1946, the first year a flu vaccine had been available. But that vaccine had been made against the previous year's flu strain—the abrupt change in the virus's genes took scientists by surprise, making their vaccine ineffective by the time people got their flu shots. There was the pandemic of 1957, the "Asian" flu that started in China and swept the world. Another vaccine failure. In 1968, there was the "Hong Kong" flu pandemic, another flu that originated in Asia. Vaccine makers were prepared, but few Americans bothered to get vaccinated. Although no epidemic even approached the 1918 one in its deadliness, Hultin could not help but worry. If vaccine makers only knew what that virus looked like, they could make a vaccine to protect people in advance and could publicize the importance of being protected. That way if or, as Hultin thought more likely, when it came again, the 1918 virus would not be so devastating.

When the 1957 pandemic struck, Hultin thought about Alaska. When the 1968 pandemic appeared, he even went so far as to contact scientists in Berkeley, California. "I thought maybe I should go back up and see if I could find more bodies," Hultin suggested. But nothing came of his inquiries and so he bided his time, waiting for science to advance to a state where it would be worthwhile for him to return to Brevig and try to find the virus that caused the 1918 flu. "I knew that sooner or later something would come up."

# 5

# SWINE FLU

On Wednesday, February 4, 1976, eighteen-year-old Private David Lewis felt feverish and achy. His nose was running, his head hurt, and he shivered with cold. When he reported for sick call, he was sent back to bed. He remained all day in a feverish haze, drifting between sleep and a sort of dazed wakefulness. But that evening, he forced himself to get up. He was a new recruit at Fort Dix in the pine barrens of central New Jersey. His unit was going out on a five-mile hike. And he was determined to join them.

As Lewis marched along, he felt worse and worse. It was becoming difficult for him to breathe—no matter how hard he tried to gasp for air, he felt he was not getting enough into his lungs. Finally, he collapsed and was rushed to the hospital, where he died a few hours later. The diagnosis: influenza complicated by pneumonia.

The first reaction at Fort Dix was shock and disbelief. Lewis was still a teenager, healthy, without a trace of chronic disease, in peak physical condition. But within a few days, Army doctors at Fort Dix and public health experts got another surprise, one that made them wonder whether Lewis's death was a perplexing tragedy or something much more ominous. Could it be, they wondered, that the unthinkable had occurred? Could Lewis's death be the first sign that the 1918 flu had returned?

The disease had begun spreading rapidly among the troops at Fort Dix the month before. Some men were taking to their beds with chills and fevers; most were simply ignoring their runny noses and malaise.

Colonel Joseph Bartley, the chief of preventive medicine at Fort Dix, was unconcerned. He was certain that the men were simply suffering from infections with an adenovirus, a benign virus that causes the common cold and can produce mild, flu-like symptoms. When Dr. Martin Goldfield, the assistant commissioner of the New Jersey Department of Public Health, suggested otherwise, saying he thought that the Fort Dix men had influenza, Bartley was so sure it was an adenovirus that he wagered with Goldfield that he was wrong. To prove it, on January 29 he sent some throat washings from ill soldiers to a medical lab for testing.

Thus began, with a casual bet and one dead soldier, what can be viewed as either a dress rehearsal for a public health crisis that never was or one of the greatest public health disasters in medical history. It was an episode that serves more as a cautionary tale than as a lesson in what not to do. For even now, more than two decades later, it is not clear that the scientists had much choice in their decisions or that, if they had to do it over again, they would make radically different decisions. The story is more nuanced, more subtle, and more difficult to dissect. But there is no doubt that it was a national testing ground and, some say, a nightmare.

It was an affair that showed how scant knowledge and real fears can be magnified in a political arena, transmogrifying into certainties that no scientist could defend and pronouncements that were based more on hype than on fact. It eventually demonstrated the unerring ability of the press to ratchet coincidences into causal relationships and spark a panic. And from its start in New Jersey to its dwindling end in the nation's courtrooms, the story was a graphic illustration of the power of an image, the haunting memory of the 1918 flu that was rising like a specter from its grave.

When Bartley made his bet with Goldfield, his reasoning that the disease at Fort Dix was caused by an adenovirus was straightforward. Men at Fort Meade in Maryland, just a couple of hours' drive away, had come down with virtually the same spectrum of symptoms as his men at Fort Dix—chills, fevers, runny noses. When medical officers sent specimens from the Fort Meade men to laboratories for analysis, it turned out that they were infected with an adenovirus. And it certainly was not surprising that an adenovirus, a type of virus that runs rampant every winter, would be spreading among the men at Fort Dix.

The winter of 1975–76 had been piercingly cold, with bitter weather that drove even the hardiest people indoors. Gray piles of hard-crusted snow cluttered parking lots and rimmed roadways. Everywhere—on buses and subways, in classrooms and offices—people were coughing and sneezing.

Fort Dix was the perfect place for an adenovirus to spread. A draft of several thousand new recruits arrived after New Year's Day, and they were joined by men just back from their Christmas break who were to be their instructors in basic training. There, in this large mixing bowl for viruses, with men arriving from across the country, it would be surprising if a respiratory virus did not take hold. Yet whatever was making the men ill did not seem particularly serious, another reason to suspect an adenovirus.

Hence, the bet.

It was resolved quickly. Bartley sent some throat swabs from sick men at Fort Dix to New Jersey's Public Health Department laboratories for analysis. A few days later, the technicians had an answer. Eleven of the nineteen samples contained an influenza virus that was infecting the populace that year, a virus known as A/Victoria, named after Victoria, Australia, where it had emerged the year before. Bartley had lost the wager.

The virology tests were perplexing. The problem was with the samples that did not contain an A/Victoria virus. While seven of them seemed to exhibit an influenza virus, the technicians in New Jersey could not figure out the flu strain. This is not, in itself, unheard of or even particularly alarming, but it did require further investigation. So Goldfield, the epidemiologist at the New Jersey lab, mailed the specimens to the Centers for Disease Control in Atlanta, a federal agency whose experts were skilled in more sophisticated viral tests and could help in puzzling situations.

In the meantime, Fort Dix soldiers continued to fall ill. When Lewis died, the doctors at Fort Dix immediately sent swabs from his throat to the New Jersey lab for analysis.

Although the virologists isolated an influenza virus from Lewis's throat sample, they could not identify it. Nor could they identify a virus from another man at Fort Dix who had gotten sick at about the same time. Goldfield sent these two samples to the Centers for Disease Control.

The day that virologists at the Centers for Disease Control received the two additional samples from New Jersey, they had completed their initial analysis of the seven New Jersey specimens that had been sent to them previously. Five turned out to contain the A/Victoria virus. But two had an influenza virus that defied their tests and required further analysis. The virus in the two samples that Goldfield sent later, including the one taken from Lewis, they discovered, was also of this mysterious strain. Four men had been infected by an influenza virus that could not immediately be identified.

It took a week for the lab at the Centers for Disease Control to discover the virus's identity. It was a swine flu virus, one that was closely related to, if not identical to, the virus that, through the sleuthing in the 1930s by people like Richard Shope, was thought to have caused the 1918 influenza pandemic. The evidence was a standard immunological test: Antibodies that cling to and inactivate swine flu viruses also clung to and inactivated this new virus. The test was to grow the virus in a fertilized hen's egg and then remove the cloudy fluid that was brimming with virus. Mix the virus with red blood cells. If they clump, you have a flu virus. Then mix the virus with antibodies to particular strains of flu and try the red blood cell test again. When you hit on the right flu strain, the virus will be inactivated by the antibodies and the red cells will no longer clump.

An antibody test was not a definite proof that the Fort Dix virus was identical to the 1918 virus. After all, no one had ever isolated the 1918 virus and the only evidence that it was a swine flu virus came from the curious appearance of swine flu antibodies in people who had survived that flu—but not in people who were born after 1918. Scientists knew that viruses that are attacked by the same antibody are either the same or closely related. And swine flu antibodies do not attach themselves to the usual influenza viruses that infect humans.

Of course, pigs had been infected with swine flu ever since 1918, but the virus had seemed to be confined to pigs. Only rarely would a pig that was ill with swine flu pass the virus on to a person, causing a mild flu. But even then, the infection would stop there. It would not spread from the infected person to other people. The strains of swine flu that had been seen before had seemed unable to cause an epidemic in humans. Nor were they deadly.

The Fort Dix situation looked very different. The men at Fort Dix did not appear to have been near pigs, which meant that a swine flu might be spreading from person to person. And one out of the four infected men had died.

On Thursday, February 12, eight days after David Lewis's death, Dr. Walter Dowdle, the laboratory chief at the Centers for

Disease Control, looked down at the written laboratory reports from the Centers' virologists telling him that one man had perished from swine flu at Fort Dix and four others were infected. He knew that this was an extraordinary moment.

It was a bit too much like the 1918 flu for comfort—a young man, healthy and strong, dying mysteriously within days of falling ill. Moreover, the men at Fort Dix were the perfect age to be infected with a flu like the one that caused the disastrous pandemic of 1918. Only people well over age fifty would have lived through the 1918 flu and could have built up antibodies that could protect them from the virus. That meant that young men, indeed most of the population, would be vulnerable if the virus came again.

Dowdle knew that federal public health officials would be at a crossroads. How could they ignore the possibility that the Fort Dix flu outbreak was a first sounding of another disastrous occurrence of the 1918 flu? If it was, they had no time to waste. In the years since 1918, medical researchers had learned to isolate and identify flu viruses and they had learned to make vaccines. Yes, it took months to make a flu vaccine. But perhaps the Fort Dix incident was a godsend, a warning that came early enough so that a 1918-type disaster might be averted. If a flu virus was going to burn through the population, leaving huge swathes of dead young people behind, it should show up as an epidemic the next fall, when new flu strains, having emerged the year before, would take over. That meant that if an unprecedented effort was made, companies might be able to manufacture enough vaccine to protect all Americans against the new swine flu.

There were a few obvious problems, including a logistical one: No one had ever attempted to vaccinate an entire population against the flu and it was hard to even imagine making so much vaccine. The other problem was scientific: The data thus far were so paltry that they could hardly justify making such a major decision.

There were no easy answers.

But Dowdle realized that this new development was so important that he could not wait for normal business hours to tell the head of the Centers, Dr. David Sencer. That evening, he called Sencer at his home, giving him the grim news. Of course, it could be a false alarm, the lab may have made a mistake, and so Sencer demanded that the virologists repeat their tests the next day.

On February 13, the virologists began to redo the tests, but it would take days to get an answer. In view of the gravity of the finding, Sencer decided not to wait. He called a meeting of federal officials for the next day, Saturday, February 14. It meant asking busy public health leaders to fly immediately to Atlanta. All agreed to come.

To be sure, the finding was ominous. Adding to Sencer's consternation was his knowledge that leading virologists, like Dr. Edwin D. Kilbourne, who was then chairman of the microbiology department at Mount Sinai School of Medicine in New York, proposed that influenza pandemics came at approximately eleven-year intervals. The virus, Kilbourne argued, periodically mutated into a new creation that few could fight off. The last pandemic was in 1968, which meant, he had predicted, that the world in 1976 was almost due for a new influenza strain. In fact, by coincidence, the same day that the virologists at the Centers for Disease Control identified the Fort Dix viruses as swine flu, *The New York Times* published an op-ed piece by Kilbourne describing the periodic nature of influenza pandemics and warning that a new pandemic was likely to strike. It appeared with a headline that said, "Flu to the Starboard! Man the Harpoon. Fill 'em with Vaccine! Get the Captain!" and it was illustrated with what Kilbourne described was a "hyperbolic cartoon," showing a man overboard grasping for a life preserver.

"I did it for the wrong reasons, in retrospect," Kilbourne said of the op-ed piece. He was becoming increasingly concerned that influenza viruses markedly changed every decade, leading to periodic pandemics. The reason, he argued, was that the virus had a limited number of proteins it could display on its surface. And the virus appeared to vary these proteins cyclically so as to maximize the likelihood that people would not have built up immunity to a strain from previous exposure to it.

History seemed to bear him out. The Asian flu virus of 1957 was thought to resemble a virus that spread through the world in 1889. The Hong Kong flu virus of 1968 was thought to resemble the virus that caused the influenza pandemic of 1898. In 1979, Kilbourne reasoned, it should not be surprising to see a flu that resembled that of 1918.

Even though Kilbourne had some reservations about predicting a massive change in the influenza virus on the basis of so few data points, he decided to go ahead. In his op-ed piece, Kilbourne stated his case succinctly, writing that influenza pandemics "have marked the end of nearly every decade since the 1940's—at intervals of exactly 11 years—1946, 1957, 1968. A perhaps simplistic reading of this immediate past tells us that 11 plus 1968 is 1979, and urgently suggests that those concerned with public health had best plan without delay for an imminent natural disaster."

Kilbourne noted that vaccines have so far failed to protect people against flu pandemics, but that did not mean that vaccines *could* not be protective. "Reasonably effective vaccines for influenza have been available for thirty years," he wrote, "but not even recent pandemics have been significantly influenced by human intervention. Whenever pandemic influenza next appears, we must improve upon our well-intentioned but uncoordinated efforts of the past that have resulted in ambiguous advice to the public and inadequate production and maldistribution of vaccine."

• • •

While Kilbourne was writing and thinking about a new pandemic, maybe even a return of the 1918 flu virus, virologists and public health officials began meeting to decide what to do about the Fort Dix virus. On the same Friday the 13th that Sencer decided to call his emergency meeting, Goldfield called Kilbourne, telling him about the Fort Dix virus and informing him that he was shipping four samples of the swine flu virus to him in New York. Goldfield knew Kilbourne well—he had been one of Kilbourne's postdoctoral fellows—so he didn't hesitate in making a request. He wanted Kilbourne to start growing the new virus in his lab, immediately, generating strains that would grow rapidly and so could be used to make a swine flu vaccine—if one became necessary. The swine flu virus that was isolated from the soldiers at Fort Dix grew poorly in the laboratory, meaning that it would have to be converted to a fast-growing strain if it was to be used for a vaccine.

Kilbourne was the world's expert at this. He had made such variants for every new flu strain that had emerged in the preceding decade. And if the swine flu virus found at Fort Dix was the first hint of a return of the 1918 flu, there was no time to waste.

"I was extremely interested," Kilbourne said. He realized that his op-ed piece with the cartoon that seemed so exaggerated at the time now looked prescient. "I thought, 'Maybe I was more right than I thought I'd be,'" Kilbourne recalled. He waited with nervous anticipation for the virus to arrive on Monday morning.

At 11 a.m., Saturday, February 14, Sencer's emergency meeting began. Secretly gathered at the Centers for Disease Control, in its collegelike campus on the outskirts of Atlanta, were the leaders of American public health. They had come to hear about the Fort Dix virus and to decide what to do.

It was an august group of men who had the power to take action—and to do so immediately, if necessary. Included were Dr. John Seal of the National Institute of Allergy and Infectious Diseases at the National Institutes of Health, whose institute studied viruses like influenza, looking at how their spread might be prevented; Dr. Harry Meyer, the head of the Bureau of Biologics at the Food and Drug Administration, which was responsible for quality control and licensing of vaccines; Goldfield of New Jersey's Department of Public Health; and two Army colonels from the Walter Reed Army Institute for Research, Philip Russel and Franklin Top, who were responsible for the health of millions of military personnel.

The men were grave and a bit on edge as the meeting began. Everyone attempted to keep the discussion focused on science. Sencer began by asking Dowdle to talk about swine flu and to describe the laboratory tests that showed a swine flu virus had infected men at Fort Dix, killing one of them. Then the officials debated what those tests might mean. Could the swine flu data be a mistake—the result of a laboratory contaminant? Goldfield said he would give new samples of the putative swine flu virus to the Centers for Disease Control for retesting in a clean laboratory, one where influenza viruses had never been studied.

But how could the extent of the swine flu epidemic—if there was one—be established? The Army said its doctors would get additional specimens from the Fort Dix soldiers who had had swine flu, and also procure blood samples from soldiers at the fort as well as their families in order to look for swine flu antibodies—evidence that individuals had been infected with a swine flu virus and recovered. The New Jersey Department of Public Health agreed to study people living near Fort Dix to see if a swine flu virus was infecting civilians in the area. Seal said that the National Institute of Allergy and Infectious Diseases would embark on a nationwide study to ascertain how widely swine flu might have spread.

Of course, everyone had one burning question: Were four cases of swine flu the first signs of a new pandemic? But, reported Dr. Arthur M. Silverstein, a doctor from the Johns Hopkins University School of Medicine who spent 1976 on the staff of the Senate Health Subcommittee, the men were nervous about the possibility, postponing an explicit discussion and playing it down, as if actually broaching the possibility of a rerun of 1918 might somehow make it happen. And yet, he added, "this idea was clearly in the minds of most of the participants." Many of the decisions they reached that day showed that they were preparing for the worst, for the return of the 1918 flu.

The group decided to make swine flu antibodies for use in laboratory tests, just in case. It is a straightforward task to produce antibodies. All you have to do is inject the virus into animals, like guinea pigs or ferrets or even chickens, and then wait a few weeks while the animals—which do not become ill or die from a human flu—make large, disease-fighting proteins. The antibodies will appear in the straw-colored blood serum, the fluid that the red blood cells float in.

The group meeting with Sencer also agreed that they needed to start preparing a vaccine. They ordered that large quantities of the newly isolated swine flu virus be produced, just in case. The Centers for Disease Control said it would start preparing special strains of the virus that would grow quickly in fertilized eggs, a necessity if the nation was to produce enough of the virus for a massive vaccination program. Goldfield then announced that he had anticipated this decision and had already sent specimens of the virus to Kilbourne. Meyer from the Food and Drug Administration said that his agency would send the special strains of flu to manufacturers as soon as they were ready so that the companies could prepare to make a swine flu vaccine.

Finally, they knew they had to do scientific studies to see whether the swine flu virus was spreading within and beyond Fort Dix and, if so, how far and how fast it was moving. The way to do this would be to take blood samples from people sick with an illness that resembled the flu, and then to take blood samples from the same people a few weeks later—the time it takes for the body to make copious amounts of antibodies. They would test people's blood for antibodies while they were ill, looking for evidence that proteins in the patients' blood could block the swine flu virus and prevent it from clumping red blood cells. Presumably, unless the people had been sick with this flu before or unless they had been around in 1918, their blood would be bereft of swine flu antibodies. But if the patients had been infected with a swine flu, their blood samples taken a few weeks after they were ill would be brimming with antibodies to this influenza strain.

The group also knew that they had no method to answer the most crucial question: Was the swine flu virus found in the Fort Dix soldiers the first sign of a recurrence of the 1918 pandemic or something totally insignificant—a feeble virus, perhaps, that could barely spread from person to person and that, even on those rare occasions when it does spread, is not harmful? There had been no deadly epidemic in the Fort Dix area yet, so uncertainty reigned. Consequently, they were worried about what to tell the public. No one wanted to start a panic, but if they kept the information secret for too long, they feared they would be savaged by the press and the public.

On Monday, February 16, Kilbourne arrived at his lab and looked for the package from Goldfield containing the viruses. It was nowhere to be found. Kilbourne was gravely concerned, worried that what might be a deadly virus had gotten lost who knows where and that perhaps by accident the vials of virus could break open and start infecting people. It turned out, however, that there was an innocent explanation. In those days, before overnight carriers like Federal Express, samples were sent via U.S. mail. That

Monday, February 16, was George Washington's Birthday, a national holiday. Kilbourne would simply have to wait another day before opening the four glass vials with black screw caps that contained the Fort Dix virus.

On February 17, the virology lab at the Centers for Disease Control reported that it had completed its retesting of the virus, confirming that the men at Fort Dix were infected with a swine flu. That same day, Kilbourne began injecting the Fort Dix virus into fertilized eggs, starting to grow strains of the virus that could be used for a vaccine.

Kilbourne worried about the danger of working with what might be a deadly flu strain. He decided that just he and his lab technician, Barbara Pokorny, would work with the viruses and that they would work in a closed room. The sort of high-tech containment facilities used today for work with deadly viruses were not available in 1976. He told no one except Pokorny what they were doing or the nature of the new viruses. Pokorny told *New York Times* science reporter Harold Schmeck months later that she kept the pact of silence. "I wouldn't let anybody in the laboratory," she said. "They really thought I had flipped out."

Meanwhile, government officials decided that the time had come to tell the public about the swine flu at Fort Dix. As much as they wanted to study the situation further and see whether a deadly flu strain was loose, they were afraid that there would be leaks to the press—the worst possible scenario.

But an announcement would have to be handled with great delicacy. No one wanted to incite a panic, especially since it was still not clear whether the swine flu was dangerous or whether it was spreading through the nation. So on February 19 Sencer called a press conference at the Centers for Disease Control. It was limited mostly to reporters in the nearby Atlanta area, but with a telephone hookup to the national press. His intention was to keep the discussion low-key and to not even mention 1918. The analogy, however, came out in the question-and-answer period that followed the official statement. A few reporters, astutely, seized upon it.

At *The New York Times*, Harold Schmeck wrote a page one story whose headline said, "U.S. Calls Flu Alert on Possible Return of Epidemic Virus." The story began by evoking 1918, saying, "The possibility was raised today that the virus that caused the greatest world epidemic of influenza in modern history—the pandemic of 1918–19—may have returned." NBC news said much the same and showed pictures from 1918 of people wearing masks in an effort to protect themselves against the killer flu.

The day after the press conference—and the very day that Schmeck's story appeared in *The New York Times*—government scientists met again, this time in Bethesda, Maryland, at the Bureau of Biologics headquarters in the sprawling Washington suburbs. Kilbourne and Albert Sabin, leading virologists from outside the walls of the government, were among those present. There was little new to say about the science—a few soldiers had been infected with a swine flu, one had died, there was no evidence of a new pandemic. Yet, said Silverstein, "for reasons that are not entirely clear, the mood seemed to have changed from 'What if . . .' to 'Well, here it is.' There appeared to be almost uniform agreement among both civilian and government scientists that the New Jersey outbreak *might* be the harbinger of more serious and widespread disease to come: while it was impossible to define precisely the extent of the risk involved, everything that they knew about influenza told them that *some* risk of disease spread existed and that it was better to be safe than sorry."

The meeting participants spoke earnestly about logistics—how to expedite the manufacture and testing of a swine flu vaccine and how to initiate a nationwide campaign aimed at immunizing virtually every American against the swine flu.

In the meantime, doctors continued to monitor the situation at Fort Dix. Men were still becoming ill with influenza, but almost all of them had the A/Victoria strain. Yet there were troubling signs that swine flu was there. Virologists found swine flu virus in a fifth man who had gotten sick in February and eight men who had recovered from flu had had swine flu, according to blood antibody tests. When doctors at Fort Dix looked for swine flu antibodies among the men, they discovered that as many as 500 of them had swine flu antibodies, indicating that they had been infected by the virus as well.

At the same time, civilians who lived near Fort Dix did not seem to be infected with anything other than A/Victoria. Neither did people living elsewhere in New Jersey. And when the Army looked at other bases they could find no swine flu. Moreover, the National Institutes of Health and state public health officials could not find any swine flu cases among civilians. When the Centers for Disease Control asked the World Health Organization to check for swine flu cases in other countries, the group reported that it could find no evidence of the virus abroad.

Goldfield, speaking about the incident a year later, was frank about the dilemma. "The experience certainly was unusual," he said. "A radically new strain had appeared in a civilian population and had died out, apparently, the first week of February. It would seem that it hadn't survived competition with A/Victoria. On the other hand, there has never been a recognition of a radically new strain of A which spread from human to human and did not turn out to be pandemic. The likelihood that the new strain could have started at Fort Dix and was found by us at the first attempt is so small that it can be dismissed as a reasonable possibility."

Kilbourne said he, too, was conflicted over the failure to find evidence that the swine flu virus had spread. "After Fort Dix, there was a long hiatus when nothing happened; in spite of the fact that there was a hothouse situation in the military barracks, it was not transmitted to the outside community."

There was a possible explanation. "The flu virus seems to disappear every spring and summer anyway. We now know that it doesn't actually disappear," Kilbourne added, explaining that flu infections are less frequent in summer but they do occur. Some people have full-blown influenza infections, with fevers and muscle aches and pains. Others have some symptoms—drippy noses, mild fevers—and often attribute them to vague entities like "summer colds." Still others are infected, and can spread the virus, but have no symptoms, Kilbourne remarked, noting that even in pandemics, at least 7 percent of people who are infected with the flu can be asymptomatic. One reason the flu is less apparent in summer is that the virus dies quickly in high humidity. It needs dry winter air to spread and flourish, which is why flu epidemics seem to disappear when spring arrives.

Even though the swine flu virus did not appear to be spreading, Kilbourne felt he could not be sanguine. The weather was getting warm, spring was approaching. "I was fearful that the virus was overwintering some place and would pop up in the fall," he says. With a vaccine program, he adds, "for the first time in history, we had a chance to forestall a pandemic."

"All of us would have liked more evidence than we have, all of us would have liked to wait," Kilbourne said in an interview at the time. "But given what could be an awesome situation, we felt we could not afford that luxury."

A year later, in an official analysis of the swine flu disaster, two medical experts, Richard Neustadt and Harvey Fineberg, wrote: "One death, thirteen sick men and up to 500 recruits who evidently had caught and resisted the disease, all in one Army camp, were the only established instances of human-to-human swine flu found around the world as February turned into March, the last month of flu season in the Northern Hemisphere."

But there was no time to waste if vaccine manufacturers were going to prepare for a swine flu immunization campaign. Normally, a group appointed by the Surgeon General—the Advisory Committee on Immunization Practice—decided in January what flu vaccines should be produced for the next year. In January 1976, the group had recommended vaccines against A/Victoria influenza, with enough vaccine to protect the 40 million Americans who were over the age of sixty-five or who had chronic diseases. By the end of February, the four companies that make flu vaccines—Merck, Merrell National, Parke-Davis, and Wyeth—had already made about 20 million doses of a vaccine against the A/Victoria flu. Clearly, that strategy would now have to be reassessed.

Kilbourne, in the meantime, had been working single-mindedly to develop a fast-growing flu strain to use for making a vaccine. It took him and his assistant two weeks to get it. They named it X-53, and although at that point, they had just a few teaspoonfuls of the virus, they began distributing it. On the weekend of February 27, the National Institutes of Health and the Centers for Disease Control sent special messengers to Kilbourne's lab to get the virus. A drug company sent a messenger to Kilbourne's home to pick up a virus sample. One week later, four vaccine manufacturers were working with Kilbourne's virus, trying to produce a swine flu vaccine.

The vaccine committee had yet to meet. Everyone realized that the group's next scheduled meeting, on March 10, was going to be seminal. The day before, Sencer got together with his staff to prepare for it. Dowdle, who was part of the group, recalled the dilemma:

"It was clear that we could not say the virus would spread. But it was clear that there had been human-to-human spread at Fort Dix. It was also clear that there was not any immunity in the population to this virus, not if you were under 50 (or maybe 62)." That meant, he said, that "most people were at risk, especially young adults. An epidemic spreading into a pandemic had to be considered *as a possibility*."

And even though the virus seemed to have disappeared from Fort Dix, after infecting only a few men there, no one could guarantee that it was really gone, Dowdle said. "Flu could do strange things. Six weeks was a short time. We had to report our fundamental belief that a pandemic was indeed a possibility."

The difficulty, Kilbourne said, was in assessing the risk. As an advisor to the proposed vaccine program and an advocate of it, he noted, "I found it difficult to convey accurately and understandably a scientifically informed perception of the relation of the new virus to the putative 1918 agent and of the hazard it presented." The Fort Dix virus came from swine, but that was a slim reed for the scientists. They did not have any way of deciding how dangerous a flu virus was without seeing it in action in a population. They could not compare the Fort Dix virus to the flu virus of 1918 because no one had any samples of the 1918 virus for comparison.

Kilbourne described the dilemma: "Therefore, one could only say that the Fort Dix virus might be more virulent, as virulent, or less virulent," in comparison to the 1918 flu virus. "The limited clinical information from authenticated Fort Dix cases was inadequate to judge the virus' potential, but its association with pneumonia and death in young recruits certainly provided no comfort."

If the government decided to go ahead with a national swine flu immunization campaign, there literally was no time to waste. It would take months to make the vaccine and eight to ten weeks to distribute it nationally, the first time ever that so many people would be receiving a vaccine. It takes two weeks for a vaccinated person to become immune to the flu. And so the time from the manufacture of a swine flu vaccine to the successful immunization of most of the nation was going to be at least three months.

One option was to make the vaccine and store it, waiting to see if a deadly swine flu pandemic really did occur. That, however, could prove disastrous, the scientists at the meeting decided, since the flu could spread throughout the world overnight. "Better to store the vaccine in people than in warehouses," one meeting participant said.

But Dowdle and others were hardly enthusiastic about taking immediate action to immunize the nation against swine flu. Neustadt and Fineberg interviewed a staff member at the Centers for Disease Control at that March 9 meeting who requested anonymity and who explained:

"There was nothing in this for the CDC except trouble. Here we were at the end of one flu season with time to try to do something before the next flu season. The obvious thing to do was to immunize everybody. But if we tried to do that, guide it, help it along, we might have to interrupt a hell of a lot of work on other diseases."

Suppose there was an influenza pandemic, the meeting participant said. An immunization program was an almost certain invitation to disaster. Those who had been unable to get the flu shots in time would be angry because they would be vulnerable. Those who were immunized but who caught another virus that they thought was the flu would be annoyed because they would assume that the vaccine did not work. All in all, millions of people would be upset. Yes, a repeat of 1918 was unlikely. But, the participant said. "who could be sure?" And if it happened, "it would wreck us."

Then take the other side of the argument, supposing there was no pandemic. Then, the staff member said, the Centers for Disease Control could be accused of wasting money, of "crying wolf," he said. Everyone, from the people who got the shots to those who administered them, would criticize the agency. "It was a no-win situation," the participant concluded.

The final decision on March 9 was predictable, however. These, after all, were people whose mission it was to protect public health and prevent disease. "Better a vaccine without an epidemic than an epidemic without a vaccine," Kilbourne said.

• • •

When the Advisory Committee on Immunization Practice convened the next day, the room was suffused with excitement and anticipation. This was to be the first public unveiling of the plan, the hoped-for campaign to immunize the nation against swine flu. The press was there, and so were experts, like Kilbourne, who had already signaled his preference that the immunization campaign go forward.

It can be difficult, more than twenty years after the fact, to know how people felt at the time. But Neustadt and Fineberg, two policy experts, interviewed the participants in the swine flu decisions a year later, while preparing a classified report—later made public—on the episode for Joseph Califano, who was the Secretary of the Department of Health, Education, and Welfare. They asked people what they thought privately and what they said publicly and tried to reconstruct a narrative of the seminal events.

In their interviews with those present at the advisory committee meeting, Neustadt and Fineberg learned that the participants had estimated, privately, what they thought the chances were of a swine flu pandemic. Estimates ranged from 2 to 20 percent, though no one discussed them. "Each was prepared to bet with nobody but himself," Neustadt and Fineberg wrote. "The probabilities, after all, were based on personal judgment, not scientific fact. They voice them to us now: they did not voice them then."

On the other hand, Kilbourne said, was it fair for the two analysts to castigate the group or to presume to read their minds? "Personally, I would have thought it absurd to quantify risk at that time beyond emphasizing 'possibility,'" he said. "Even if the odds were guessed to be 100 to 1 against occurrence of a pandemic, the *gravity* of the risk was high."

Neustadt and Ernest R. May, a historian at Harvard University, later analyzed what they saw as crucial moments in the swine flu vaccine decision. It shared features, they said, with several other crucial moments in history, the Bay of Pigs and the Vietnam War. In each instance, they say, "the individuals who made the key decisions, or at least some of them, looked back and asked, 'How in God's name did we come to do *that*?'"

That March 10 meeting of the vaccine committee was a turning point, illustrating how the specter of 1918 overwhelmed even the experts in the field.

Neustadt and May take special note of the fact that the experts at the March 10 meeting never publicly stated what they thought the chances were that the United States would be swept by a deadly swine flu epidemic if the nation took no action to immunize the population. That, they say, was a major mistake. Forcing experts to give odds can be one of the best methods for exposing fundamental weaknesses in an argument, they note. Certainly, they argue, before making a major decision, like the decision to start a swine flu immunization campaign, federal officials should have asked the medical experts to openly state their best guess of the chance that a deadly epidemic would occur. "If the doctors and their fellow experts hesitate, we offer the suggestion of one academic colleague with extensive government experience," they say. "Ask instead, 'When I brief the press, with you by my side as an expert, and tell them the odds are X, will I be right? No? Then how about Y? And so forth.'

"Once differing odds have been quoted, the question 'why?' can follow any number of tracks. Arguments may pit common sense against common sense or analogy against analogy. What is important is that the expert's basis for linking 'if' with 'then' gets exposed in the hearing of other experts before the lay official has to say yes or no."

Another way to expose hidden assumptions and unstated value judgments is to pose what Neustadt and May call an "Alexander's question," named after one of the men at the March 10 meeting, Dr. Russell Alexander, a public health professor at the University of Washington. Alexander posed a question to the group that Neustadt and May thought so apt that, they say, if policymakers routinely asked an "Alexander's question," historic blunders, as well as commonplace blunders, might be avoided.

Alexander's question was brilliantly simple. He asked what information might make the group change its mind about the need to prepare to immunize the nation against swine flu? Would it be evidence that every swine flu case was mild? Or that no one but the Fort Dix soldiers got the swine flu? Would it make a difference what the timing of the outbreaks was or where they occurred?

When Alexander asked his question at the meeting, Neustadt and May note, he "never got an answer. In the circumstances, it was the right question," they add. "Pursuing it would have flushed out a deeper set of questions, which also did not get asked, questions about tradeoffs between side effects and flu, questions about programming and scheduled review, questions distinguishing severity from spread, questions about stockpiling, and more."

In fact, Neustadt and May realized, "What 'Alexander's question' forces into the light are causal associations thought to be validated by past experience." It would have showed the power of the analogy of the 1918 flu and exposed the paucity of scientific data behind the swine flu decision.

Despite the way his question was avoided, Alexander nonetheless urged caution and suggested that it might be better to make the vaccine and hold it until it was clear that there was a dangerous pandemic underway. But Alexander was a quiet man who did not press his points. He gave no speeches and said relatively little. In fact, noted Neustadt and Fineberg, Alexander appeared "unimpassioned" and "so mild that other members we have seen recall but vaguely something about 'stockpiling.' He himself makes light of it. Known as a voice of caution in past meetings, he was easy to discount on this occasion."

When Neustadt and Fineberg interviewed Alexander in preparation for their report on the swine flu affair, he told them, "My view is that you should be conservative about putting foreign material into the human body. That's always true . . . especially when you are talking about 200 million bodies. The need should be estimated conservatively. If you don't need to give it, don't."

Alexander's worries were easily overlooked in the zeal of the moment to implement a historic public health effort, to immunize a nation against what could be a deadly pandemic. And the arguments of those who wanted to go ahead were hard to ignore.

Seal later recalled that one staff executive from the Centers for Disease Control privately told Sencer, "Suppose there is a pandemic accompanied by deaths. Then it comes out: 'they had an opportunity to save life; they made the vaccine, they put it in the refrigerator . . .' That translates to 'they did nothing.' And worse, 'they didn't even recommend an immunization campaign to the Secretary.'"

For some, the urge to go forward went beyond a desire to protect the nation from disease and extended to a desire to demonstrate the importance of the very field of public health at a time when it was seen as less glitzy, less interesting than the burgeoning area of molecular biology. As Dr. Reuel Stallones, who was dean of the Public Health School at the University of Texas, observed, "This was an opportunity to try to pay something back to society for the good life I've had as a public health doctor. Society has done a lot for me—this is sheer do-goodism. It was also an opportunity to strike a blow for epidemiology in the interest of humanity. The rewards have gone overwhelmingly to molecular biology, which doesn't do much for humanity. Epidemiology ranks low in the hierarchy—in the pecking order, the rewards system. Yet it holds the key to reducing lots of human suffering."

Several of the scientists were careful to express reservations, however, noting that it was entirely possible that there would be no swine flu epidemic the next winter, making a vaccine campaign unnecessary. Nonetheless, Kilbourne said, "our reservations, though voiced, were subservient to our mutual wish that the program proceed."

Looking back on those days, Kilbourne asked himself why he and others did not push harder for stockpiling the vaccine, an option that, in retrospect, "seems so obvious and attractive." He had to remind himself of two pressing reasons that persuaded him that the best choice was to vaccinate everyone:

First, he said, if the group had decided to make and store the vaccine, that would have led to "faltering Congressional support for what was already a troubled program. Momentum would have been lost."

And second, Kilbourne recalled, "those responsible for the vaccine program in the field told us that the problems of distribution, setting up clinics, and vaccination itself would preclude prompt action if rapid spread of the virus occurred in the customary winter season."

By early afternoon of March 10, the group had reached an agreement—to go ahead with a national campaign to immunize all Americans against swine flu.

"Stallones summed it up best," noted Sencer. "*First* there was evidence of a new strain with man-to-man transmission. *Second*, always before when a new strain was found there was a subsequent pandemic. And *third*, for the first time, there was both knowledge and the time to provide for mass immunization." So, he said, "if we believe in preventive medicine we have no choice."

Sencer wrote a nine-page memo, which became known as his "action memorandum." His goal now was to convince the government to start the program. Although he included several options, starting with "do nothing," he laid out the reasoning behind the advisory committee's consensus, mentioning specifically the possibility that the 1918 flu could return, and recommended that the federal government buy the vaccine in sufficient quantities to immunize all Americans, that the National Institute of Allergy and Infectious Diseases conduct field trials, that the Bureau of Biologics then license the vaccine, that a mixture of public and private groups do the actual immunizations, and that the Centers for Disease Control do surveillance. The cost of the vaccine would be $100 million, Sencer wrote. The rest of the program would cost an additional $34 million. "We have not undertaken a health program of this scope and intensity before in our history," he explained. "There are no precedents, nor mechanisms in place that are suited to an endeavor of this magnitude."

Sencer wrote the memo for his boss, Dr. Theodore Cooper, the Assistant Secretary for Health, and addressed it to David Mathews, who was Secretary of the Department of Health, Education, and Welfare. In the end, it went all the way up the federal chain of command, landing in the hands of President Gerald Ford, and it became the paper that drove the decision.

Neustadt and Fineberg observe that the memo "reads as though it were deliberately designed to force a favorable response from a beset Administration that could not afford to turn it down and then to have it leak." It had just that effect.

Sencer wrote his memo on Saturday, March 13. On Monday, March 15, he was in Washington, meeting with Mathews, trying to sell it. Sencer's appointment with Mathews followed immediately after Mathews's daily staff meeting. That day, Dr. James Dickson, the Deputy Assistant Secretary for Health, was present and he described the situation as Sencer had laid it out. Inevitably, the discussion centered on the possibility that the 1918 flu was about to return.

Dickson said later that it was inevitable that Mathews would want to move forward with an immunization campaign. "I presented the issue to Mathews . . . He said to me, 'What's the probability?' I said, 'Unknown.' From the look on Mathews' face when I said that, you could take for granted that this decision was going to be made."

Mathews agreed. "The moment I heard Sencer and Dickson, I *knew* the 'political system' would *have* to offer some response," he said, adding that "it was inevitable."

He reasoned that if the probability that the 1918 flu was coming back was "unknown," that meant that the chance was greater than zero. And that was enough to force the issue. "You can't face the electorate later, if it eventuates, and say well, the probability was low so we decided not to try, just 2 to 5 percent, you know, so why spend the money," he said.

That morning, Mathews wrote to James T. Lynn, director of the Office of Management and Budget, explaining that he was recommending the swine flu campaign: "The indication is that we will see a return of the 1918 flu virus that is the most virulent form of the flu. In 1918 half a million people died [meaning Americans]. The projections are that this virus will kill one million Americans in 1976."

Within five days of the meeting of the vaccine advisory committee, the odds of a catastrophic flu epidemic had risen to a near-certainty. People at that meeting said there was a "possibility" of a pandemic. But, Neustadt and Fineberg observed, "Sencer's action memorandum three days later used the term *strong possibility*. Now Mathews had changed the 'possible' into a 'will be.'" And the Fort Dix virus, described at the vaccine meeting and by Sencer as resembling the 1918 virus because both viruses were grasped by swine flu antibodies (hardly proof that the viruses were identical), became, in Mathews's words, "a return of the 1918 flu virus." Moreover, Neustadt and Fineberg noted, "although the scientists repeatedly said that they had no way of assessing the virulence or severity of the new virus, Mathews extrapolated the half million deaths in 1918 to a projection of one million deaths in 1976, since the population had doubled."

That same day, President Gerald R. Ford sat down to a meeting with a few of his staff members—James Lynn, the budget office director, Paul O'Neill, Lynn's deputy, and James Cavanaugh, the deputy director of the Domestic Council. Swine flu was not on the agenda, but the three men brought it up and told Ford that an immunization campaign would require supplemental funds. A week later, Ford heard about it again, in a formally scheduled session, allotting a half hour for what was to be a full review of the subject

The room was crowded with a familiar cast of characters, as well as a sprinkling of new faces to the debate. Mathews was there, as were Theodore Cooper, who was Sencer's boss, Lynn and O'Neill. But this time Earl Butz, the Secretary of Agriculture, was also present, telling Ford that although an unprecedented number of eggs would be needed to grow the flu virus in order to make the vaccine, "the roosters of America are ready to do their duty."

Among the papers handed out at the meeting was Sencer's memo, with its alarming predictions and persuasive appeal. Some warned Ford that no matter what he did, he might be criticized. Sencer's memo was looking more and more like a gun to the heads of the policy advisors. If the government decided to do nothing and that memo got out, it could be a nightmare, and in a presidential election year.

Hearing only what might go wrong if he failed to authorize a campaign and not what might go wrong if an immunization project went forward, Ford came to what was probably a preordained conclusion. He explained: "I think you ought to gamble on the side of caution. I would always rather be ahead of the curve than behind it. I had a lot of confidence in Ted Cooper and Dave Mathews. They had kept me informed from the time this was discovered. Now Ted Cooper was advocating an early start on immunization, as fast as we could go, especially in children and old people. So that was what we ought to do, unless there were some major technical objection."

Ford's next step was to call a meeting with leading scientists, giving his decision the imprimatur of the most impressive medical researchers. It was held on March 24, at 2:30 in the afternoon, in the Cabinet Room of the White House. Those attending included Kilbourne, Stallones, and Jonas Salk and Albert Sabin, two feuding doctors who were American heroes for their roles in wiping out polio with a vaccine.

Some of the specialists in influenza saw the new flu strain and its ties to the 1918 flu as an excuse to try an immunization campaign that might vastly improve upon what happened in the last flu pandemic, the Hong Kong flu of 1968. That time, too few were immunized, and too late, to staunch the virus's spread. But most others had a different response to the 1918 metaphor.

It "came as a bolt from the blue," to many who had to make a decision, Neustadt and May wrote, "capturing imaginations and dominating impressions. Though the 1918 influenza holds but a small place in most histories, biographies, and memoirs, it seems that almost everyone at higher levels in the federal government in 1976 had a parent, uncle, aunt, cousin, or at least a family friend who had told lurid tales of personal experience with the 1918 flu. The killer had been known then as 'Spanish flu': the term 'swine flu' meant nothing much to laymen off the farm. But the year 1918—more precisely 1918–1919—cited in conjunction with the flu called up vivid images in Washington almost sixty years later. Those images were rooted in folk history and were more powerful because of it."

When Sencer and the leading scientists of the day met with President Ford, their discussions were overlaid with the metaphor of the 1918 flu. And that metaphor drove the decision.

Sencer opened the meeting by reviewing the facts of the potential swine flu epidemic, as he saw them. Ford then asked Salk and Sabin for their opinion. Both were enthusiastically in favor of a swine flu campaign. Finally, Ford asked those who wanted the nation to proceed with a swine flu immunization effort to raise their hands. All did.

With the scientists' explicit recommendation that the swine flu vaccine program proceed, Ford dismissed the meeting and said he would adjourn to the Oval Office. Anyone who wanted to speak to him privately there was urged to do so. "Just get up, come over, knock, and walk in," Ford told them. No one did.

Ford felt confident that the medical community was behind the decision to start a swine flu campaign. But, Neustadt and May argue, that consensus was less than firm. Look at the medical experts who were invited to the meeting. Salk and Sabin agreed with each other for once and supported Sencer and Cooper, who were sponsoring the program. Sencer himself had suggested the others at the meeting, and most of them had already made up their minds to support the program. Alexander was there but, Neustadt and May observed, he "rarely spoke up anyway." The others "were already committed to Sencer's plan." As a result, the group's "unanimity meant less than Ford assumed. It neither kept Sabin and Salk together (Sabin turned against the program three months later) nor reflected firm views in the medical community where opposition (and indifference) mounted as months passed without swine flu," they write.

Ford was unaware of these complications. When no one crept into his office to express reservations about the program, he concluded that there were no doubters. So he decided to go ahead, reasoning, as he said, that "if you've got unanimity, you'd better go with it."

Ford strolled into the Cabinet Room and asked Salk and Sabin to join him. Then he walked over to the press room to make his announcement, telling the nation about an unprecedented effort that was about to take place to stop a deadly flu.

Flanked by Sabin and Salk, Ford began: "I have been advised that there is a very real possibility that unless we take effective counteractions, there could be an epidemic of this disastrous disease next fall and winter here in the United States." Ford said: "Let me state clearly at this time: no one knows exactly how serious this threat could be. Nevertheless, we cannot afford to take a chance with the health of our nation."

With that preamble, Ford announced that he was asking Congress to appropriate $135 million "for the production of sufficient vaccine to inoculate every man, woman, and child in the United States," for a disease that no one could even prove to exist.

# 6

# A LITIGATION NIGHTMARE

It should have been a scene of triumph when President Gerald Ford appeared before the nation to announce his swine flu vaccine. All the elements were there: Advances in science and medicine were going to allow humans to take arms against the virus. The most revered doctors in America, Dr. Jonas Salk and Dr. Albert Sabin, were giving the battle their blessing, flanking President Ford in solidarity. And the end result was going to be an unprecedented effort to prevent what might be a return of one of the worst plagues ever to strike humanity in recorded history.

But the trouble began almost immediately.

The very day that President Ford made his announcement that he was seeking $135 million for a national swine flu vaccine program, critics who had kept mum until that crucial moment suddenly surfaced.

Maybe it was the sheer hubris of President Ford and the two doctor-gods of vaccines, thinking that they could pull this off, assuming that the press would remain credulous and simply report the news. Maybe it was that just beneath the surface of the seemingly unified federal bureaucracies lay dissenters, political advisors, and scientists who told themselves that even if the decision makers did not want to hear them, the public would. Or maybe it was just that there was a critical mass of people who were convinced this swine flu immunization campaign was a cockeyed idea. For whatever reason, the critics lay in wait, ready to spring up and wag their nagging fingers of doubt, under the cloak of anonymity, if they could.

Reporters found these skeptics, of course. It is part of a reporter's job to ask who, if anyone, disagrees with a pronouncement, especially one as dramatic as Ford's decision to start the swine flu immunization campaign.

Two CBS correspondents began by making the rounds of Ford's political advisors, asking them questions about the real reasons for a swine flu vaccine program. Was it, they asked, a political move designed to bolster Ford's popularity? Was it meant to counteract Ford's reputation for being weak and indecisive, a bumbler? The advisors said that they themselves were skeptical of the campaign; as a group they conveyed a distinct lack of enthusiasm.

The reporters were not just selectively hearing the grumblers either. In a detailed official investigation of the swine flu affair a year later, Richard E. Neustadt and Harvey Fineberg interviewed all the participants and verified for themselves the same surprising lack of support for the program. They report that they "ranged across the list of Ford's political advisors, covering them thoroughly, we believe, from top to bottom, without finding an enthusiast among them."

The reporters also learned that scientists within the Centers for Disease Control were saying privately that the national immunization campaign was unjustified and even crazy.

It was a reporter's bonanza. Here was the President, and the most revered doctors in America, stating categorically that a massive immunization campaign could save the nation from a looming plague. And here were the true experts, those silent minions who gave political and scientific advice, saying that the whole idea was loony, misguided, a political ploy, scientific idiocy.

With their eye-opening reporting in hand, the reporters were prepared. All they needed was a decision by President Ford to go ahead with his program. Then they could let loose and tell the nation that there was by no means a consensus that such a program was wise.

And so, on March 24, 1976, on the *CBS Evening News* with Walter Cronkite, the very news program that announced President Gerald Ford's decision to start the swine flu immunization program, one of the reporters, Robert Pierpoint, said: "Some experts seriously question whether it is logistically possible to inoculate two hundred million Americans by next fall. But beyond that, some doctors and public health officials have told CBS News that they believe that such a massive program is premature and unwise, that there is not enough proof of the need for it, and it won't prevent more common types of flu. But because President Ford and others are endorsing the program, those who oppose it privately are afraid to say so in public."

As so often happens in politics—and journalism—once the door was cracked open for critics to enter, others followed. The next day, Dr. Sidney Wolfe, director of the Health Research Group, a Ralph Nader organization in Washington, came forward, echoing the concerns that Pierpoint raised.

The stage was set for a rancorous debate. Before the vaccination program even began, before there was any money appropriated for the swine flu program, the sniping had started. Worse was yet to come. Something that had only been hinted at, as a sort of hypothetical possibility, but never taken seriously by policymakers or scientists, would turn out to be the first true nightmare of the immunization campaign.

• • •

But, for the moment, Ford's program went forward. After all, even though a substantial number of prominent critics had qualms, it was, in the end, hard to argue that it was better to do nothing, or to stockpile a vaccine, with the shadow of another 1918 flu epidemic looming.

Geoffrey Edsall, a former professor of public health at Harvard University, happened to be at the Bureau of Biologics of the Food and Drug Administration when the decision was made to ask Ford to start the swine flu program. "When one of them told me they would recommend a national vaccination program, I questioned this on the basis of the slim evidence at hand," Edsall said. "He replied, 'Look, I know that the chance of a pandemic may be as little as 1 to 50 or even less, but if *you* were the President of the United States and were told that the country faced a 1 to 50 or maybe even a 1 to 100 chance of a national disaster—which on the basis of all available evidence could largely be averted by a vaccine program—what would you say?' I got the point."

So did Congress, which immediately approved the full $135 million that Ford requested to immunize virtually the entire nation. Although the Senate and the House of Representatives held hearings in which they questioned the gravity of the swine flu threat, the hearings turned out to be pro forma. Dr. Theodore Cooper, the Assistant Secretary for Health, testified both times that the threat was real—the 1918 flu might be coming back—and that his goal was to inoculate 95 percent of the population, 200 million people, a goal that had never been achieved before with any immunization campaign.

Cooper ran the program through the Department of Health, Education, and Welfare, delegating responsibility for various parts of it to different federal agencies. For example, the Centers for Disease Control was responsible for monitoring the nascent swine flu epidemic and attempts to quell it. So the agency set up a surveillance system that would track swine flu cases throughout the country, record the progress of the immunization campaign, and look for adverse reactions to the vaccine.

The Bureau of Biologics coordinated actual vaccine production, making sure that there was a good supply of a fast-growing swine flu virus and that there would be tens of millions of fertilized hen's eggs available in order to grow the virus for the vaccine. In addition, vaccine makers were told to stop making a vaccine to protect against what everyone had thought would be the predominant flu strain in the fall, A/Victoria, and shift over to making only swine flu vaccine. The vaccine makers had already produced about 30 to 40 million doses of A/Victoria vaccine, but that would not go to waste, it was decided. It would be mixed with swine flu vaccine and given to people at high risk of becoming seriously ill or dying from the flu—mainly the elderly.

But as the program got going, the smallest details became issues, even the very name of the disease. Pig farmers complained to the Centers for Disease Control that the name "swine flu" might frighten people away from eating pork. They asked, to no avail, that the flu's name be changed to "New Jersey flu."

More serious controversy among scientists continued, and this time it took place in public. At a meeting held at the Centers for Disease Control on April 2, Dr. Martin Goldfield, who was New Jersey's chief epidemiologist, and among the first to know about the Fort Dix swine flu problem, told the assembled bureaucrats and the press that he thought the swine flu campaign was a bad idea and that healthy people might experience serious side effects from the vaccine. That night, on the *CBS Evening News*, he was featured saying, "There are as many dangers to going ahead with immunizing the population as there are with withholding. We can soberly estimate that approximately fifteen percent of the entire population will suffer disability reaction."

*The New York Times* published a series of editorials, by Harry Schwartz, that castigated the swine flu program, arguing that it was scientifically unjustified and that the vaccine might prove dangerous.

At first, the critics had little effect on the advice given to the public. The Department of Health, Education, and Welfare, for example, analyzed what newspapers across the country were saying in their editorial pages. It looked at eighty newspapers from sixty cites and determined that, as of April 2, 1976, 88 percent had supported the swine flu program.

Some scientists, including Dr. Edwin Kilbourne, the virologist who had been a key advisor to the vaccine program, wrote letters to the editor of *The New York Times* taking issue with Schwartz's editorials. But increasingly other scientists were starting to say publicly that they thought the swine flu immunization effort was ill-advised. Even Dr. Albert Sabin of polio fame, who was last seen standing at President Ford's side as he announced the swine flu campaign, began questioning it now that it was underway. He spoke at the University of Toledo on May 17, recommending that the vaccine be made but held in storehouses unless and until it was clear that a deadly swine flu epidemic was in progress. And Dr. J. Anthony Morris, an administrator at the Bureau of Biologics, the very agency of the Food and Drug Administration that evaluates vaccines and rules on their safety, began speaking out against the program. He said that it would be dangerous to give the vaccine to so many and that the vaccine was likely to be ineffective in any case.

With the steady drip of criticism, the tide started to turn, as reflected by the views of newspapers that had initially been supportive of the immunization program. When the Department of Health, Education, and Welfare surveyed the nation's newspapers in May, it found that only 66 percent were still promoting the swine flu program, a big drop from the 88 percent that supported it a month earlier.

On June 2, the vaccination program got more bad news. Parke-Davis had made several million doses of vaccine, but had mixed up its viruses. Instead of making vaccine against the swine flu strain found in New Jersey, it had made vaccine against the strain that Dr. Richard Shope had isolated from pigs forty years earlier. The company would have to throw away all of that vaccine and start over again using the correct virus. Other companies had their own problems making enough vaccine when they discovered that they were getting lower vaccine yields than they expected—one dose of vaccine per egg instead of two.

Outside the United States, other nations looked on with detachment. Some could not afford a campaign like the one underway in America; others were wary and decided to simply stockpile the vaccine and possibly provide it to people at high risk of serious illness if they were infected with the flu. There were a few countries, like the Netherlands, that decided to follow the U.S. lead and immunize, but they were the exceptions. Dr. Nancy Cox, who is current chief of the influenza branch at the Centers for Disease Control and Prevention, as the agency is now called, says that it is not surprising that the United States was virtually alone in the world in reacting so strongly to the threat of a deadly influenza epidemic. The seminal events—the emergence of the swine flu and the death of a young soldier—took place on American soil. "That always makes a country much more likely to act," she said. Moreover, Cox added, even though many countries had an air of detachment, that does not mean they were not concerned. "Many throughout the world were extremely worried and were watching," to see if the Fort Dix virus was going to spread and be a killer.

Yet, some argued, the evidence was going the other way, starting to show that the Fort Dix virus might be a dud. In July, in an experiment that still dismays some virologists, British scientists inoculated six volunteers with the New Jersey swine flu virus. Five got only a mild disease and the sixth had no signs of illness. The scientists and volunteers were lucky, Kilbourne noted. "It has always seemed to me that the investigators took a shocking risk in inoculating humans with a virus of unknown potential—indeed with the potential of starting an epidemic," he said. But the study was not criticized as unethical; rather it was cited as a reason to doubt the wisdom of a vaccine program. Britain's leading influenza expert, Sir Charles Stuart-Harris, said that "it is indeed highly questionable whether the amount of vaccine required for those between 20 and 50 years of age should be prepared at the present time for any country, including even the United States, until the shape of things to come can be seen more clearly."

The next blow was the results of the field tests of the swine flu vaccine—the attempts to see if volunteers who were given the vaccine developed antibodies that would protect them against the swine flu. The vaccine worked well in people over age twenty-four—those who received it developed abundant antibodies against the swine flu. But children were not well protected, which meant that much more vaccine would have to be produced and that children would have to be induced to return for a second vaccine shot, complicating efforts to immunize the population. Yet it was crucial to vaccinate children, who might be expected to spread the flu in schools and day-care centers, sparking an epidemic.

In a season of bad surprises, none surpassed the next to beset the swine flu program. The vaccine makers announced that they could not get liability insurance for the vaccine. And until they could, they said, they would not bottle it.

The four vaccine manufacturers had hinted all along that insurance could be an issue, but those concerns were more or less swept aside by federal officials concentrating on the logistics of getting a vaccine program of unprecedented scope underway in record time. The American Insurance Association and individual insurance companies had argued that the federal government should indemnify the vaccine makers, but few in the government took them seriously. The thought was that they were bluffing. Federal officials were convinced that all that would be necessary to satisfy the insurance industry would be for the government to assure them that it would be responsible for warning against possible side effects and for making sure that those vaccinated gave their informed consent.

That idea was nonsense, the insurers retorted. The more they thought about the liability problem, the worse it seemed.

At one insurance company, whose officials spoke to historians of the episode under condition of anonymity, the problem surfaced when some junior underwriters brought the problem to the management's attention. The company's president says that the underwriters focused on their own personal responsibility if the company were hit with big claims from people who said the vaccine injured them. "Their worry," the president said, "was, 'how would a catastrophic award look on my record?'" But once the issue was raised, more and more people at the company got concerned. And, the executive says, as the liability question worked its way up through the management ranks, "the higher it got, the more it was seen in the broader management context, claims and suits: incalculable amounts of overhead cost for which we had unlimited liability. Our vice president . . . decided not to insure. Then he put his decision to me, which he usually wouldn't do, because of the White House announcement, the public affairs aspect. I simply confirmed his decision . . . If the public really was endangered, the government should take the risk; it certainly could, we wouldn't."

Until President Ford launched his swine flu campaign, few outside of drug companies had thought much about who was liable when someone claimed injury from a vaccine. There had, however, been one notable vaccine lawsuit, and it had sent chills through the industry, making the companies wary of what might happen with a swine flu vaccination campaign.

The litigation involved a polio vaccine, made by Wyeth Laboratories, and it had concluded in 1974, just two years before the swine flu program was announced. The case, known as *Reyes v. Wyeth Laboratories, Inc.*, involved an eight-month-old baby who got polio after receiving a Wyeth vaccine. Experts testified that the company had done nothing wrong and that the baby's case of polio was probably not related to the vaccine. But Wyeth lost and was ordered to pay $200,000 to the child's family. Wyeth appealed to the Supreme Court, but the Court refused to hear the case and the judgment against the company held.

Neustadt and Fineberg, in their postmortem on the swine flu affair, tartly summed up what the Reyes case meant to vaccine manufacturers. The courts had ruled that Wyeth had failed to adequately warn of its vaccine's dangers. "Never mind that the company had included in cartons for shipment a printed form which *did* contain adequate warning. Never mind that experts had testified at trial that this particular case was not vaccine-related. Wyeth would pay (and did). The suffering was real and Wyeth had the only deep pocket available."

What might happen if an entire nation got a flu vaccine? By chance alone, tens of thousands of people who were vaccinated would come down with some illness after getting a flu shot and some would die. After all, tens of thousands of people get sick and many die every day. That is simply the normal course of events. But what if some of these people claimed that the vaccine made them sick, or led to the deaths of loved ones? What would a jury say if confronted, for example, with a child who began having severe epileptic seizures within hours or days of being immunized against swine flu? How sympathetic might a jury be to a large company that argued that epilepsy was coincidental and not related to the vaccine when they saw the tearful parents and the concerned doctors who argued the opposite? Or how about the young woman who got multiple sclerosis after her flu shot? Or the middle-aged man who suffered a heart attack or stroke? Then multiply those cases by tens of thousands, and companies could envision financial catastrophe.

Even worse, what if the vaccine really did cause a disease, albeit one that was not known when the vaccine campaign began? How well would companies fare if plaintiffs' lawyers argued that the companies should have known, or that they actually did know about the association between the vaccine and the disease—that obscure and unpublished data buried in their files actually told them about it but they hid those findings and hoped no one would ever discover them?

It simply was not worth taking a chance, the companies reasoned. No matter if they warned of any and all known dangers of the vaccine. No matter if they could easily explain to the scientific community the illnesses and deaths that were bound to occur by coincidence when more than 100 million people were vaccinated. The fact remained that the companies could be sued, and they could lose big. Even if they won, they could be saddled with immense costs of defending themselves against a barrage of lawsuits.

Dr. Hans H. Neumann, who was director of preventive medicine at the New Haven Department of Health, explained the problem in a letter to *The New York Times*. He wrote that if Americans have flu shots in the numbers predicted, as many as 2,300 will have strokes and 7,000 will have heart attacks within two days of being immunized. "Why? Because that is the number statistically expected, flu shots or no flu shots," he wrote. "Yet can one expect a person who received a flu shot at noon and who that same night had a stroke not to associate somehow the two in his mind? Post hoc, ergo propter hoc," he added.

In addition, Neumann wrote in his letter, within a week after receiving flu shots, 45 people will develop encephalitis and more than 9,000 will get pneumonia. Nine hundred will die of pneumonia. "Sequential to the immunizations? Yes, but not a consequence of them," he said. "These are only a few examples of what is bound to happen the day and the week after immunizations."

Neumann cautioned: "It is one thing to see matters objectively in light of statistical expectations. It is quite another when it affects one personally. Who can blame someone for assuming the events are linked? Hence, the presumption of tort and the liability problem that is expected."

Even in the Department of Health, Education, and Welfare, which wanted to indemnify the companies, there was dissension in the ranks. One staff member, who worked for Cooper, the head of that department, told Neustadt and Fineberg why he and others objected to having the government rather than the vaccine makers be held liable for vaccine-related injuries: "Behind these arguments for indemnification, there were a number of assumptions which were untested and unsupported by facts. For one, it was contended that if the manufacturers were not indemnified, they would all stop making vaccine. But the number of companies in this business had been diminishing for a long time for reasons totally unrelated to liability. We just couldn't buy this—that continued liability would drive them out. And then there were other unsupported assumptions, just sort of out there, loping across the plains."

As the vaccine makers met repeatedly with lawyers for the Department of Health, Education, and Welfare, they became increasingly frustrated. One company lawyer explained: "We would open every meeting with a heartfelt refrain for the HEW lawyers: 'We need legislative relief. Nothing short of that is going to do it. Chairman Rogers [Paul G. Rogers, chairman of the House Subcommittee on Health and the Environment] is willing to put it in a bill. We need legislative relief.' That was our first paragraph at every session. It fell on absolutely deaf ears."

Inevitably, vaccine production was delayed.

On May 21, a leading vaccine maker, Merrell National, told the chief lawyer for the Department of Health, Education, and Welfare that the company would not provide the swine flu vaccine unless the federal government provided indemnification. On June 10, the insurers for Parke-Davis and for Merrell National told the companies that their liability coverage for the swine flu vaccine would expire on July 1. The only solution would be for Congress to pass a law requiring the federal government to insure the vaccine makers.

On July 15, Merrell said it would be stopping its vaccine production entirely and that it would not even purchase eggs after July 20.

Congress held hearings. The insurance companies did not budge. They simply could not assume the risk for the vaccine makers, they insisted.

The vaccine manufacturers, in the meantime, were making the swine flu vaccine in bulk but were not putting it in vials so it could be distributed. It would take weeks to package the vaccine, delaying even further the start of an immunization campaign that now was beginning to seem hopelessly mired down.

The impasse lasted until August 1, when a swine flu scare spurred Congress to act.

At an American Legion convention in a Philadelphia hotel a group of people fell ill and twenty-six died of a mysterious disease. It seemed to be a respiratory disease. It looked, in fact, like the flu, and some doctors said publicly that the men might have died from swine flu. For four days, while television stations showed funerals of the Legionnaires and the new disease made headlines, it seemed that the predicted flu epidemic had begun.

On August 5, the Centers for Disease Control completed its laboratory studies of the disease. Whatever was sickening these men, the data showed, it was not swine flu. (Later, the culprit turned out be a hitherto unknown bacterium that had gotten into the hotel's air-conditioning system and was pumped throughout the building.) But even though Legionnaires' disease, as the illness became known, was not swine flu, the message was not lost on Congress: if it had been swine flu instead, the criticisms of Congress would have been withering and the ensuing panic impossible to counter with arguments about liability insurance. If it turned out that the American people were denied a vaccine because Congress refused to give legal protection to the vaccine makers, it could be a political nightmare. So Congress acted quickly, passing a "tort claims bill" that required that any claims arising from the swine flu vaccine be filed against the federal government. The bill, which came before the Senate on August 10, was rushed through without hearings or a committee report. The next day, it went to the House of Representatives, where it passed even though many members had not seen the legislation.

Senator Harrison A. Williams, Jr., of New Jersey, said that the law broke new ground. "This is pioneering, in a sense," he said. But, he added, "it is in response to an emergency."

In the House of Representatives, Paul G. Rogers of Florida urged Congress to step in to help the vaccine makers. The federal government, he explained, has "asked the drug companies to produce this vaccine. We have told them how to do it. We have told them the dosage we want, what strength. We gave them the specifications because we are the only buyers, the Government of the United States. This is not the usual process of going out and selling." But "if someone is hurt, we think people ought to have a remedy."

On August 12, President Ford signed a bill into law committing the federal government to insuring the swine flu vaccine makers against claims that their product injured people.

A Gallup poll taken on August 31 found that 95 percent of Americans had heard of the swine flu vaccination program and that 53 percent planned to take part. Although officials at the

Men line up in Love Field, Texas, in 1918 to have their throats sprayed with antiseptic as a preventive measure against the flu *(Courtesy of the National Archives, 165-WW 269 B-36)*

An influenza ward in U.S. General Hospital #16 in New Haven, Connecticut, in 1918. Sheets were hung between the beds in an attempt to contain the virus (*Courtesy of the National Archives, 165-WW 269 B-40*)

Public health departments distributed masks during the pandemic in an effort to prevent the spread of the virus. Here the 39th Regiment, on its way to France, marches through the streets of Seattle, Washington, wearing masks provided by the American Red Cross (*Courtesy of the National Archives, 165-WW 269 B-8*)

A man is turned away by a Seattle trolley conductor because he is not wearing a mask (*Courtesy of the National Archives, 165-WW 269 B-11*)

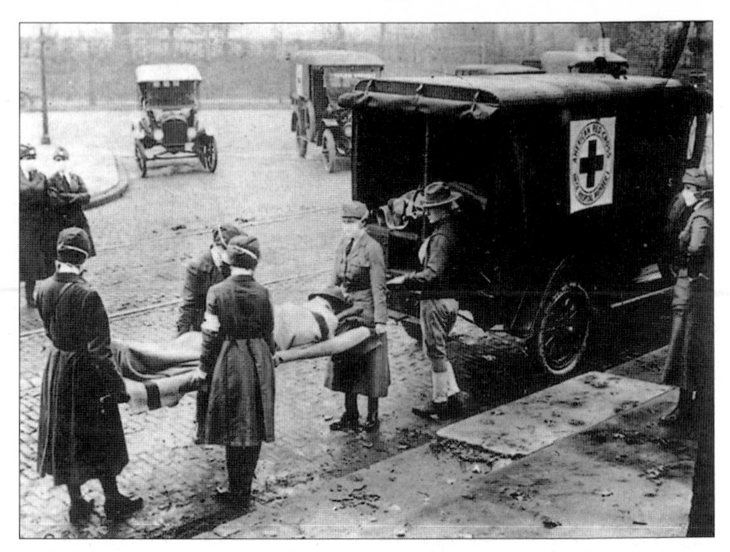

In St. Louis, Missouri, the Red Cross Motor Corps transports a flu victim *(Courtesy of the National Archives, 165-WW 269-B-3)*

An unidentified minor league baseball game in 1918 *(Courtesy of Stanley B. Burns, M.D., and The Burns Archive)*

Johan Hultin and his colleagues in Brevig, Alaska, in June 1951, standing in the mass grave of flu victims whose bodies had been preserved by permafrost since 1918. From left, Hultin, Otto Geist, Jack Layton and Albert McKee (*Courtesy of Johan Hultin*)

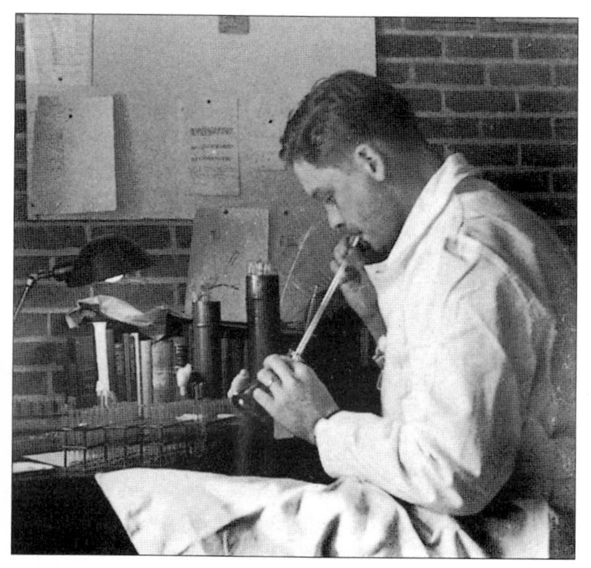

Johan Hultin working in the Microbiology Department at the University of Iowa in 1951. He hoped to grow the virus in the amniotic fluid of fertilized chicken eggs (*Courtesy of Johan Hultin*)

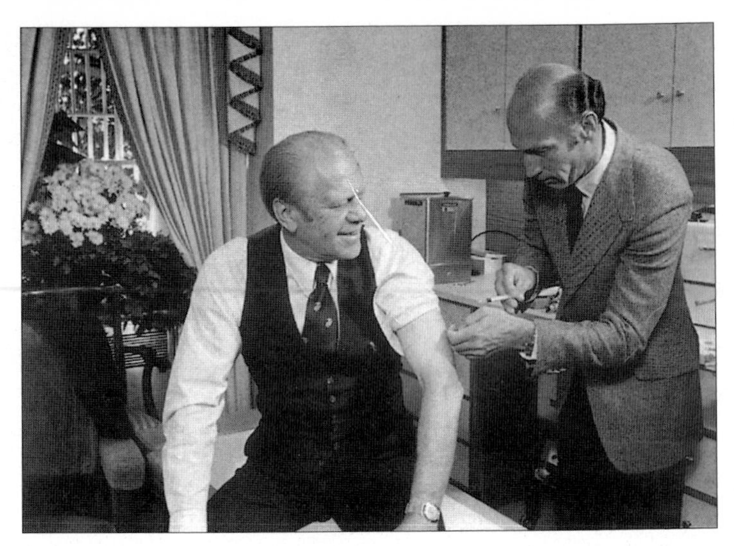

In 1976, afraid that the 1918 virus had reappeared in the form of swine flu, the Federal government instituted a national immunization campaign. When some who had been vaccinated died, President Gerald Ford was immunized in an effort to assuage public fears. Here he receives his flu shot from Dr. William Lukash *(Courtesy of the Gerald R. Ford Library)*

People lining up for flu shots in New Jersey during the 1976 influenza immunization campaign *(Courtesy of the Centers for Disease Control)*

In August 1997, Johan Hultin returned to the mass grave in Brevig, Alaska, to look for lung tissue. He found the corpse of a woman whose lungs had been preserved since 1918. Here he crouches in the grave beside the woman's remains *(Courtesy of Johan Hultin)*

Jeffery Taubenberger and Ann Reid examine DNA readouts at the Armed Forces Institute of Pathology in Washington, D.C. They are studying tissue samples from victims of the 1918 flu to try to determine what made the virus so deadly *(Courtesy of Eric Haase)*

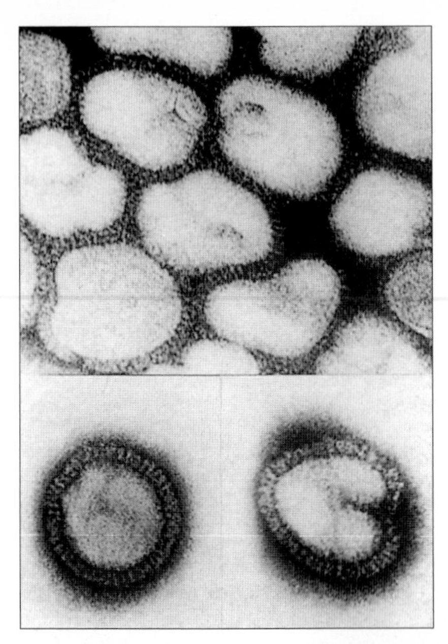

Electron micrographs of the influenza A/PR/8/34 virus, which was isolated in Puerto Rico in 1934 (*Courtesy of M-T. Hsu and Peter Palese*)

Lung tissue samples, preserved in paraffin, from victims of the 1918 flu. These and more than 3 million other tissue samples are stored in the National Tissue Repository maintained by the Armed Forces Institute of Pathology (*Courtesy of Eric Haase*)

Centers for Disease Control were disappointed—they were aiming for 95 percent participation—the poll nonetheless showed that the message had gotten through. An ominous flu could be on its way, a repeat of 1918's epidemic, and the government was going to sponsor an unprecedented immunization program to protect Americans.

The first Americans were immunized on October 1. Ten days later, the first deaths occurred.

Three elderly people in Pittsburgh, all of whom had heart problems, died suddenly after receiving swine flu shots. All had gotten their shots at the same clinic. All had had shots from the same batch of vaccine. And, reported the *Pittsburgh Press*, that same batch of vaccine had been delivered to twelve other clinics in Allegheny County as well as to clinics in twenty cities across the country.

The media swarmed to the Pittsburgh clinic, and the next day, October 12, Dr. Cyril Wecht, the Allegheny County coroner, fanned the flames of fear, going before CBS television cameras and saying that a bad batch of vaccine "is definitely a possibility that must be considered."

Allegheny County suspended its swine flu campaign. And so did nine states. The press began a national body count.

Some newspapers went over the top. The *New York Post*, for example, ran a story on October 14 headlined "The Scene at the Pennsylvania Death Clinic" that spoke of a seventy-five-year-old woman who "winced at the sting of the hypodermic," then had "taken a few feeble steps and dropped dead." On October 25, the paper suggested that Carlo Gambino, the mobster, had been killed by the Mafia using a swine flu shot as the deadly weapon.

As reports of the death toll proliferated, Dr. David Sencer, at the Centers for Disease Control, tried to stem the rising tide of fear. He held a press conference on the evening of October 12, saying that there was no evidence that the vaccine was at fault. The deaths were most likely only coincidentally associated with the vaccination. But the government would, of course, investigate. "We are setting up a program to look into this in great depth to reassure everyone that this is not a problem due to the vaccine but just some of the inherent problems of providing preventative services to large numbers of people, particularly those who are elderly and have underlying health problems."

It was the start of a tug-of-war between the Centers for Disease Control and the Pennsylvania coroner.

The next day, October 13, Wecht announced that autopsies on two of the three people had shown that they had died of heart problems. But, he said, the vaccine may have spurred those deaths. "We know that substances injected into the vascular system directly produce a more exaggerated and certainly a more rapid reaction than when those same substances are injected into the body fat or muscle mass."

The Centers for Disease Control countered with its own figures on the likelihood of coincidental deaths, noting that among people aged seventy to seventy-four, there are 10 to 12 deaths per 100,000 people per day. So, of course, when you immunize people of that age, some, by chance, would die the same day. But that does not mean they died because they were immunized.

Some medical experts were acutely aware that spurious and potentially alarming associations between the vaccine and deaths would be sure to crop up if they were not extremely careful. Dr. Robert B. Couch, who is professor and chairman of the department of microbiology and immunology at Baylor College of Medicine in Houston, said he and his colleagues went to a large nursing home near the medical college to immunize the residents with test batches of the swine flu vaccine. "We were welcomed; they opened the doors," he recalled. "But then we learned they had an average of one death every two days." If they immunized the residents, Couch realized, "no matter what we did we'd have a relationship between the flu vaccine and deaths." So did he go ahead and immunize them anyway? "No, we did not," Couch replied. "We thanked them very much," he said, but "we chose to leave that one."

In the meantime, however, the public was growing increasingly jittery. To soothe their fears, President Ford and his family got their flu shots on October 14, and made sure that they were televised doing so. Still, the newspapers continued their body count. The *New York Post*, in an article that same day that featured a photograph of Edwin Kilbourne, the flu expert, getting his flu shot, announced that so far thirty-three people had died in sixteen states. No matter how passionate health officials were in their attempts to reassure the public, the Pittsburgh scare and its aftermath were impossible to shake. Polls showed fewer and fewer Americans planned to be immunized.

Nonetheless, by mid-December, 40 million Americans, a third of the adult population, had had swine flu shots. It was twice as many as ever before immunized against flu in any single season and it was the largest vaccination program in history. All the while, however, a disaster was brewing.

In the third week of November, a doctor in Minnesota called his local health officials to report that he had immunized one of his patients against swine flu and the man had developed a nerve disorder called Guillain-Barré syndrome. It is a rare disease, afflicting about 4,000 to 5,000 Americans each year, and its cause is unknown. It begins with a tingling in the hands and feet and weakness in the arms and legs. The disease can progress and affect nerves that control breathing and swallowing. Within one to two weeks, the symptoms reach a crescendo and then gradually diminish over a period of weeks or months. Most people fully recover, but about 5 percent die from respiratory problems and about 10 percent are left with some degree of permanent paralysis.

The Minnesota doctor whose patient developed Guillain-Barré syndrome said he was on the lookout for such an effect. He had heard a medical education tape, he said, that warned of Guillain-Barré syndrome as a possible side effect of the swine flu vaccine.

A Minnesota health official, Denton R. Peterson, told the Centers for Disease Control about the case, but encountered little interest from the federal officials. Peterson was worried. He told Neustadt and Fineberg, "We felt we were sitting on a time bomb." Several more people in Minnesota developed Guillain-Barré syndrome after getting flu shots and one died. Peterson called the Centers for Disease Control again. By this time, three Guillain-Barré cases had surfaced in Alabama, and the next day, November 20, there was a case in New Jersey.

At the Centers for Disease Control, Sencer's staff began searching the medical journals, asking whether there was any hint in the published literature of a link between flu vaccines and Guillain-Barré syndrome. Were the cases something to be expected, a normal, though rare, consequence of all flu vaccines? Or was there something peculiar about this flu vaccine? Or were they simply a statistical fluke, a chance occurrence that had nothing to do with the fact that the patients had just been immunized against swine flu?

A search of the published papers was reassuring. The Centers for Disease Control announced to the public in December 1976 that they had found reports of 1,100 cases of Guillain-Barré syndrome. One person had had a flu shot before becoming ill. But four had been struck by lightning before developing the disease. The published reports on Guillain-Barré hardly hinted at an association with flu vaccines, the Centers for Disease Control concluded.

But that, of course, did not reveal whether *this* flu vaccine, the swine flu vaccine, might be causing the disease. In order to answer that question, epidemiologists at the Centers for Disease Control would have to know how common the disease was, which would allow them to say whether there had been a sudden spike in cases following the swine flu immunization campaign. On November 20, the very day that the Centers for Disease Control learned about the New Jersey Guillain-Barré patient, Dr. Leonard Kurland, a renowned neurologist and epidemiologist at the Mayo Clinic, got a call from the agency asking whether he had any information on the number of people who would be expected to get the disease, vaccine or no vaccine.

Kurland was happy to help. He replied that, by chance, the Mayo Clinic had just finished investigating that very question. The clinic keeps medical records of all the residents of Olmsted County, Minnesota, where the clinic is located. Looking through those records from 1935 through 1968, Kurland and other researchers at the Mayo Clinic had discovered 29 cases of Guillain-Barré syndrome, giving a rate of about three cases for every 200,000 people per year.

The next question was more troubling: Had Kurland seen any patients recently who had had swine flu shots and then developed Guillain-Barré syndrome? Kurland, unaware of the report by Denton Peterson, was puzzled. Why, he wondered, was he being asked? The reason, he was told, was simply that the Centers were doing a surveillance study. But the message was not lost on Kurland. "Of course you come away with the idea that there was some association," he said.

Kurland thought it made eminent sense that there might be a relationship. After all, Guillain-Barré syndrome was a peculiar disease. It was rare—so uncommon that only his study group, which specialized in such projects, had ever carefully assessed how often it occurred. It was unusual. And its cause was unknown. But if there was a surge in the number of people getting this disease after they had received a swine flu shot, he was certainly not going to argue with the hypothesis. He himself had not seen any Guillain-Barré patients who had just gotten a swine flu shot, but others, of course, had.

At the same time, however, Kurland was concerned that the data could quickly become muddied if it became known that an association between the swine flu shot and the disease was expected. He warned Dr. Philip Brotman, the Centers for Disease Control scientist who was in charge of keeping watch for Guillain-Barré cases, that his task was tricky. "It is a very tough diagnosis to make. You must have a neurologist present and at that time we did not have a substantive set of diagnostic criteria," Kurland explained. The criteria were mostly descriptive: neurologists would be looking for a progressive nerve disease that started in the feet and moved upward through the body, causing weaknesses as nerves failed to fire properly and so muscles that depended on those nerves could not contract.

Kurland said he tried to caution Brotman to be careful about publicizing his investigation of the putative link between Guillain-Barré syndrome and the vaccine. Guillain-Barré syndrome, he emphasized, is just one of several disorders that can cause symptoms like nerve weakness and diminished reflexes. He feared that if it was widely known that the Centers for Disease Control was looking for an association, doctors might make a diagnosis of Guillain-Barré syndrome immediately if a vaccinated person came in but might reach a different conclusion if someone else appeared with the same symptoms but who had not received the swine flu vaccine. In other words, the very fact that the Centers was looking for an association might bias the reporting and make an association appear.

"Let's say you're a neurologist and two patients come to see you with weakness in the legs and no apparent reason," Kurland said. "There are many, many causes for weakness," he added. But with the putative swine flu association, a neurologist will naturally ask the patients if they had a swine flu shot. Suppose one did and one did not. For the patient who had the shot, the doctor will immediately suspect Guillain-Barré syndrome, Kurland asserted, and the doctor would be likely to seize on that as an explanation for the person's symptoms.

The other patient, who did not have a swine flu shot, poses more of a diagnostic problem. "You are much less sure of what the patient may have," Kurland said. "There are a hundred different ways you could sign it out." And so, he said, those who were not vaccinated are much less likely to be diagnosed as having Guillain-Barré syndrome just as those who were vaccinated were much more likely to be told they have the disease.

Kurland believes his concerns were borne out. Within a few days after the Centers for Disease Control called him, he heard from other neurologists at the Mayo Clinic who had also been notified. "By this time, with the publicity, a vaccinated person, I believe, was far more likely to be labeled GBS in the sign-out diagnosis," he said.

Calling on neurologists in eleven states to report all new cases of Guillain-Barré, the Centers for Disease Control soon got what it feared: doctors from across the country started sending in reports of recently immunized patients who developed the nerve disease.

The reports were ominous, some medical experts said. Few people with Guillain-Barré syndrome became ill in the week after they were vaccinated. Most got sick two to three weeks after their flu shot and few got the neurological illness four or more weeks after their inoculation.

Dr. Lawrence Schonberger, a young medical epidemiologist at the Centers for Disease Control, was put in charge of analyzing the data. The more data he got, the worse it was starting to look and the more scientists and administrators at the agency began to wonder what they should do. After all, the Centers for Disease Control had started this major immunization campaign and now it looked like not only was there no danger from swine flu but the vaccine itself might be injuring people.

"I would present what I was doing at higher and higher levels at the CDC," Schonberger said. "It peaked when I was talking to the director of the CDC." On a conference call while Schonberger sat in the office of David Sencer, the director, were leading academic experts who were helping Sencer decide what to do. The decision was that the association was not quite pinned down. On Wednesday, December 15, the Centers notified doctors across the nation that they should continue giving swine flu shots.

Schonberger felt that he just hadn't made himself clear. "It was visual, I could see the visual and say there's a pattern here. I wasn't sure I could communicate that," he said. He woke up from his sleep at two o'clock the next morning, shouting to his wife, "Rachel—I've got it." He suddenly knew how to analyze and present the data so that everyone would see why he was convinced that the vaccine was causally linked to Guillain-Barré syndrome. That morning, he rushed to work and began redoing his analysis. A couple of additional cases of Guillain-Barré syndrome had been reported by then, making the association stronger. This time he was persuasive. On Thursday, December 16, Sencer conceded that the swine flu vaccination program must be stopped because of the possibility that the vaccine might be causing Guillain-Barré syndrome.

That day, with President Ford's agreement, Dr. Theodore Cooper, Assistant Secretary of the Department of Health, Education, and Welfare, announced that the swine flu immunization campaign was over. There had been not a single case of swine flu and the prospect of any danger at all from the vaccine was chilling.

The press was not kind. Harry Schwartz, the *New York Times* editorial writer, termed the swine flu campaign a "fiasco." Writing on December 21, he blamed, among other things, "the self-interest of the government health bureaucracy which saw in the swine flu threat the ideal chance to impress the nation."

Reports of Guillain-Barré cases continued to pour in to the Centers for Disease Control. But even as the toll of the injured and dead mounted, some scientists echoed the concerns that Kurland had voiced early on. The problem, some neurologists told the federal scientists, was that the disease was poorly described and so doctors could include a variety of symptoms under the rubric "Guillain-Barré." At a meeting on December 29, Dr. Dale McFarlin, who was chief of the neuroimmunology branch at the National Institute of Neurological and Communicative Diseases, spelled out his concern. "I think until you have some hard-nosed criteria," he said, "these data don't really mean very much."

But, at a time when ironies and mistaken assumptions were making a shambles of the swine flu vaccine, the secret irony of the Guillain-Barré link was perhaps the wryest one. The whole alarm came about because the Minnesota doctor who first reported a flu shot patient who got the disease had misheard an audiotape. He thought the tape warned that Guillain-Barré syndrome might follow flu shots. In fact it said just the opposite. It had used the disease as an example of how a faulty association could be drawn between a disease and the vaccine if, by chance, a person who had had the vaccine happened to become ill for reasons having nothing to do with the vaccine.

The tape, unearthed by Neustadt and Fineberg in their investigation of the swine flu affair, was of a lecture by Dr. Paul F. Wehrle, who was speaking at a conference at the University of California in Los Angeles. He said: "Problems with diseases that may be confused or incorrectly interpreted as induced by influenza vaccine I think will occupy a lot of our time and a lot of our attention. We have at any one time in the state of California somebody who is in the process of developing the Landry-Guillain-Barré-Strohl or Guillain-Barré syndrome, the paralytic episode with sensory loss and so on, just spontaneously. Usually we have no indication of anything that's setting this kind of problem off. I can assure you, though, that if someone is developing this and receives the influenza vaccine within about thirty days of the time of onset of that illness, this influenza vaccine will be blamed either for initiating it or making it worse."

The aftermath of the fleeting campaign to immunize all Americans against the swine flu was a tidal wave of litigation. People with Guillain-Barré syndrome were demanding compensation. So were people with a myriad of other diseases—multiple sclerosis, rheumatoid arthritis, polymyositis, fainting spells. Some reported that the vaccine gave them flu-like illnesses causing them to miss days at work. Others claimed they had heart attacks or strokes, even impotence. All said they had become ill after receiving a swine flu shot. All wanted the government to pay. The government, after all, had said it would be responsible for compensating people who had been injured by the swine flu vaccine. And now here they were—the injured.

The centerpiece of the litigation hinged on a study by the Centers for Disease Control, which reported on 1,098 cases of Guillain-Barré syndrome reported by state and territorial health officers. The risk of getting Guillain-Barré syndrome appeared in the first six weeks after people got a flu shot, peaking in the third week. The Centers concluded that the vaccine increased people's risk of getting the disease eightfold, and that the risk period extended for ten weeks after the immunization.

The Centers also turned up people with a variety of other neurological disorders that commenced after the flu shots: facial paralysis, nerve inflammation, encephalitis injuries to nerves of the hands or feet, brachial nerve inflammation, inflammations of the optic nerve, people whose nerves were losing their insulating layer, called myelin.

The Guillain-Barré cases, however, stood out, triggering the law requiring that the federal government compensate vaccine victims. The decision was to compensate anyone who got the disease within ten weeks of a swine flu shot. The cases were to be tried by federal district judges, without a jury, and so it would be up to the judges to decide who had a legitimate claim and who did not.

The law allowed people to file claims within two years after their injuries, and the claims steadily dribbled in. In September 1977, 743 claims had been filed, seeking $325,671,708 in damages. There were also 67 claims of wrongful death, including 19 of death from Guillain-Barré syndrome, and these claims sought an additional $1,032,948,179. (This included a claim for $1 billion that was subsequently withdrawn.) Neil R. Peterson, a Justice Department lawyer who gave an accounting of the claims to Congress, said at the time that about 20 a week were coming in and that he expected that the final tally would be 2,500 claims. But by May 1980, there had been 3,917 claims filed, seeking more than $3.5 billion in damages.

For judges, the lawsuits were an ordeal. These were not clear-cut cases and they did not involve a well-defined disease. Instead, as doctor after doctor testified in court, Guillain-Barré disease was poorly defined and its cause was unknown. Its symptoms could be anything from mild weakness in the legs to total paralysis. There was no definitive laboratory test for the disease, although there was at least a common clinical sign: doctors have found increased levels of protein in patients' cerebrospinal fluid, levels that start to rise a few days after symptoms appear and continue to increase for the next four to six weeks. But there can be other reasons for increased amounts of protein in a person's cerebrospinal fluid. And not everyone who claimed they had Guillain-Barré syndrome had had this test; some who had had it did not have increased protein levels.

Some medical experts thought that Guillain-Barré syndrome could be triggered by a viral illness—colds and flu or diarrhea, in particular—or a vaccination, like the swine flu shots. The idea was that the body would make antibodies in response to the disease or the vaccination. Some of those antibodies could, by accident, start to attack and destroy the fatty sheath of insulation that surrounds nerves. The result could be Guillain-Barré syndrome. Nonetheless, the hypotheses were speculative.

The medical mysteries swirling around the disease were irrelevant to the litigation, however. The only requirement for people seeking compensation was that they show that their complaints were caused by the disease and that they developed the disease within weeks of being immunized.

It was not as straightforward as it sounded, however, to determine who, of the thousands of plaintiffs, should be compensated.

The story of the swine flu litigation played out in the courtroom of one federal judge in particular, Sherman Finesilver, of Denver, who found himself responsible for hundreds of cases and who decided that he needed help to cope with them.

The difficulty, as Finesilver immediately realized, was that the swine flu cases were not clearly defined situations in which a young, robustly healthy person had a swine flu shot and then, shortly afterward, developed Guillain-Barré syndrome. There was an elderly woman with a checkered medical history. A man who got Guillain-Barré syndrome so long after the flu shot that it was hard to argue cause and effect. Nothing was simple. Yet Finesilver found himself called upon to decide between conflicting testimony from medical experts and to determine which expert to believe. "I had to ferret out where the causal connection was," Finesilver said.

He found himself asked to make decisions that depended on two areas he had avoided in his academic career: statistics and medicine. "I became a lawyer because I didn't like science or medicine," Finesilver said. But faced with the swine flu vaccine cases, he learned.

Finesilver enrolled in neurology courses at the University of Colorado's Health Sciences Center in Denver. He was an odd sight in the classroom, dressed in a suit, when the hundred-odd young medical students were attired in jeans, looking like the "rejects from a foreign army," Finesilver recalled. "They stared at me—this guy in the back of the room taking notes." But he absorbed the technical details. "To this day, I can tell you paragraph 56 in the neuropathy syndrome" in the neurology textbook, he said.

The problems were evident from the very first swine flu case to reach Finesilver's courtroom. The patient was Jennie Alvarez, a sixty-three-year-old woman who lived in a small farming town near Greeley, Colorado. She had worked at a variety of jobs until her retirement in 1975; her most recent one was housekeeper at the county hospital.

On October 28, 1976, Jennie Alvarez had a swine flu shot. Three weeks later, she said she began feeling tired and drained of energy; she complained of shooting pains in her shoulders, elbows, wrists, and knees. Seven months after her flu shot, Alvarez was hospitalized. The diagnosis: Guillain-Barré syndrome. Two years later, when she finally had her day in court, her legs were paralyzed and she used a wheelchair.

On the face of it, it sounded simple. Cause and effect. But the more that was known about Jennie Alvarez's medical history, the less clear her story became.

Jennie Alvarez had had a complicated medical story, dating back to 1968, when she was diagnosed with arthritis. It became so severe that she was hospitalized in April 1976, and her medical record stated that she "had arthritis all over and could hardly get around and that she walked with a marked stoop and very slowly because of the pain and discomfort."

But arthritis was not her only medical problem. She also had intestinal ailments, such as ulcers, gastritis, colitis, and diverticulosis, as well as arthritis in her neck and shoulders. At times she complained of aches in her legs and hips. In October 1975 and in April 1976, her doctor, Robert Porter, diagnosed Alvarez with "nervous exhaustion."

Alvarez, however, argued that her flu shot threw her into a downward spiral of increasing fatigue and debilitation that culminated with Guillain-Barré syndrome. She said that as the weeks and months passed after she had been immunized against swine flu, she became progressively weaker. Although she had had medical complaints, she had once been a vigorous woman, she testified. In the spring of 1976, for example, she planted a vegetable garden. That summer, she said, she painted her kitchen and tiled her bathroom floor.

Alvarez said that her problems began three weeks after she had her swine flu shot. She became bone-weary, so exhausted that she could not even cook Thanksgiving dinner and so tired that she sent her family home early on Christmas Day, asking them to leave right after dinner because of her fatigue. In those dreary months that followed her flu shot, Alvarez said, she found herself lying down in the middle of the day, needing to take a rest. When she went to church, she dared not kneel because she could not raise herself when she did. On top of the fatigue, Alvarez said she was plagued with sharp pains in her shoulder, elbows, wrists, and knees.

Alvarez's lawyers argued that her symptoms were caused by a mild form of Guillain-Barré syndrome, a subclinical disease that was triggered by her flu shot and that "smoldered" for seven months. It ignited in May 1977 when she developed gastroenteritis, Alvarez's lawyers said. That illness caused her subclinical

Guillain-Barré disease to erupt, violently attacking her nerves and resulting in a characteristic, full-blown case of Guillain-Barré syndrome.

Lawyers for the federal government argued that Alvarez was suffering from progressive arthritis, which explained her weakness and aches and pains. Her Guillain-Barré syndrome, diagnosed in May, was caused solely by her gastrointestinal disease, they contended.

Eight doctors testified at Alvarez's trial. Testifying for Alvarez were Dr. Martin Lewis, who was professor and chairman of the department of neurology at Georgetown University, Dr. Charles Poser, a professor of neurology at the University of Vermont, Dr. Sidney Duman, a neurologist in Denver, and Dr. Peter Quintero, also a Denver neurologist. Testifying for the government were Dr. Barry Arnason, professor and chairman of the department of neurology at the University of Chicago, Dr. James Austin, professor and chairman of the department of neurology at the University of Colorado Medical Center, Dr. Stuart Schneck, a professor of neurology and neuropathy at the University of Colorado Medical Center, and Dr. Lawrence Schonberger, the epidemiologist at the Centers for Disease Control whose data analysis convinced the government to stop the swine flu immunization campaign.

Those who believed Alvarez was a swine flu vaccine victim subscribed to the smoldering-illness hypothesis, saying that her flu shot stimulated her immune system, priming her body to respond with Guillain-Barré syndrome when she developed diarrhea, nausea, and vomiting in May 1977. The experts did not agree on all of the details of their hypothesis. But, as Finesilver wrote in his opinion: "We are dealing with a constantly evolving, uncertain medical science. While Plaintiff's theories diverge at certain points, there is a common thread among them, i.e., the swine flu vaccination produced a malfunction in the immune system which caused it to react improperly to the gastroenteritis."

The experts testifying for the government argued that the swine flu shot had nothing to do with Alvarez's medical problems. Her complaints, they said, stemmed from her arthritis and not from a disease of her nerves. Yes, they said, she did apparently develop Guillain-Barré syndrome in May 1977. But, they added, that illness was most likely precipitated by the virus or bacterium that caused her gastrointestinal illness. Although the cause of Guillain-Barré syndrome is not known, these experts argued that it sometimes follows an infection or a vaccine injection. And when that happens, they said, the syndrome occurs in weeks—not seven months—after the triggering event. There is no such thing, they said, as a "smoldering" Guillain-Barré syndrome.

Finesilver was not persuaded by Alvarez's medical experts. He wrote: "It is unfortunate that Mrs. Alvarez suffers from this enigmatic disorder. She has demonstrated throughout her lifetime attributes of hard work, frugality, and family concern. However, the fact remains that only those illnesses which are proven to be causally related to the vaccine are compensable." And Alvarez, Finesilver concluded, "has failed to establish by a preponderance of the evidence that the GBS from which she suffers was caused by the immunization."

Faced with 126 cases from six states, Finesilver did not relish his task of deciding who should be compensated and who should not. He realized that rule 706 of the Federal Rules of Evidence, which set up standards for judges to use in their courtrooms, allowed judges to appoint panels of scientific experts who would work independently of the plaintiffs and the defense and try to sort out where the truth lies. Finesilver decided to appoint a panel of three medical experts to help with the swine flu vaccine cases. "I was the first to use rule 706," Finesilver said.

The panel members, Dr. Stanley Appel, a neurologist at Baylor College of Medicine in Houston, Kurland of the Mayo Clinic, and Dr. Clark Millikan, a neurologist at the University of

Utah School of Medicine in Salt Lake City, agreed to conduct medical exams of the plaintiffs, review all documents submitted by lawyers on either side, and decide whether the plaintiffs' injuries could have been caused by their swine flu shot.

It worked, Finesilver says. Of the 126 cases, he settled all but four out of court.

While the lawsuits poured in to the federal government, Leonard Kurland found himself unable to forget the swine flu vaccine problem. The more he investigated, the more he doubted whether the vaccine ever had anything to do with Guillain-Barré syndrome, becoming convinced that the entire association was spurious, a consequence of biased reporting.

The first hint came when Kurland saw data from the armed forces, where 80 percent of the 2 million people on active duty were immunized—a total of 1,700,000 people. In fact, Kurland said, they each got double doses. The military, he says, "wasn't taking any chances." The question was: Did Guillain-Barré syndrome strike these people after they had swine flu shots? The military doctors were careful in diagnosing the disease, sending people to hospitals where they were seen by neurologists. Yet, Kurland notes, there was no association between the swine flu vaccine and Guillain-Barré syndrome in these military personnel. "We found 13 cases in the three services just after the vaccine," Kurland said. Then, he said, he and his colleagues looked at the number of cases in previous years to see how many would be expected. The number? Seventeen.

Then there was the experience in the Netherlands, which had jumped on the swine flu bandwagon, unlike virtually any other country, and vaccinated more than 1.5 million people. There was no increase in the numbers of people getting Guillain-Barré syndrome.

Kurland turned to the Mayo Clinic data, which contained the medical records of all residents of Olmsted County. Forty thousand people got the vaccine; no one had a well-documented case of Guillain-Barré syndrome occurring within weeks of the flu shot. There was one case, but it was hardly what a specialist would call documented, Kurland notes.

The woman, Kurland says, was not seen by a neurologist but instead was examined by a family doctor. The documentation consisted of a note stating that the patient was an eighty-year-old woman who had had weakness in her legs, and since she had had a swine flu shot, it was most likely that she had Guillain-Barré syndrome. "There wasn't enough evidence to call that a case," Kurland remarked.

So why did the Centers for Disease Control find an association between the swine flu vaccine and Guillain-Barré syndrome? The answer was clear, Kurland asserted. It was the old problem that he had worried about from the start—the problem of the self-fulfilling prophecy.

The Centers for Disease Control, Kurland discovered, did not have a set of specific tests and symptoms to define Guillain-Barré syndrome. And, he said, the agency did not obtain copies of clinical records for review. Moreover, there was no follow-up of cases, even though Guillain-Barré syndrome can only be diagnosed with certainty when doctors have a complete record that shows how the patient's symptoms progressed.

Instead, Kurland says, when the Centers for Disease Control called for data, "they had students or whoever could get to the facilities in the states go out and try to find cases. Anything that was brought in with Guillain-Barré mentioned was put down as a potential case." At the time, doctors were already aware that the Centers suspected that the flu shots might cause Guillain-Barré and so, Kurland notes, a bias was introduced.

Kurland recalls the eighty-year-old Mayo Clinic woman with the weakness in her legs. "To make a diagnosis on that basis was shocking," he remarked. Yet, he fears, that was what happened when the Centers for Disease Control put out a call for patients who had swine flu shots and then developed Guillain-Barré syndrome.

Kurland was and remains a world-renowned expert on epidemiology and neurology, noted Robert B. Couch, the flu expert at Baylor College of Medicine. Moreover, Couch said, no matter what you think of the initial report by the Centers for Disease Control citing a link between the swine flu vaccine and Guillain-Barré syndrome, there have been about a half dozen studies on the relationship over the years and one thing is certain. "There certainly is an inconsistency between swine flu vaccine and Guillain-Barré syndrome," Couch said. "That certainly should cast doubt on the relationship."

Lawrence Schonberger of the Centers for Disease Control disagrees. His original observation stood the strictest test, he said, one that Kurland himself participated in. As the dispute continued, a group of leading researchers, including Kurland, tried to resolve it by examining the actual medical records of each patient with Guillain-Barré syndrome in Michigan and Minnesota during the period from October 1, 1976, to January 31, 1977. They found that those who were immunized against swine flu were seven times more likely to develop the neurological disease than those who did not have a swine flu shot. While Kurland says the data were tainted by doctors' assumptions that anyone who had a flu shot and reported neurological symptoms was likely to have Guillain-Barré syndrome, he nonetheless signed off on the paper.

"That's what he reported," Schonberger said. "It was his study. Not my study. His."

The military data and the data from the Netherlands are flawed, Schonberger added. In the Netherlands, the population was small and the number of cases of Guillain-Barré syndrome was correspondingly small. Unconvinced that the swine flu vaccine posed a danger, the Dutch actually used it for two years. In the first year,

Schonberger said, there was a slight increase in the number of Guillain-Barré cases. There was no increase in the second year, but the U.S. experience indicated that *any* previous flu shot made Guillain-Barré syndrome less likely to emerge after a swine flu shot. It was only when the Netherlands data for both years of swine flu shots were lumped together that the association went away.

As for the military data, there were several problems, Schonberger said. The period of time when the Guillain-Barré cases were counted was three to four months, while the disease is most likely to occur within six weeks of a vaccination. With a longer time period, the link between the flu shot and the neurological disease gets washed out, he said. In addition, the military population was younger, and younger people were at less risk of getting Guillain-Barré.

Epidemiologists were torn. They called upon the competing camps to debate their evidence at closed meetings of the American Epidemiological Society, setting up what were essentially battles of the titans. One of the first was at a meeting in Philadelphia, before an audience of about 200 leading experts. On the stage were Kurland and Colonel James W. Kirkpatrick, the chief of preventive medicine for the Army, facing off against Dr. Alexander Langmuir, a legendary Harvard epidemiologist, and Schonberger.

"They raised their points and we raised ours," Kurland said. The heated debate went on for nearly an hour.

Schonberger was the Young Turk, seated among the giants and taking part in a dispute that he could never have imagined, witnessing Kurland and Langmuir disagreeing so heatedly. Both, he said, were icons. "It was really quite a scene. Here were two people who were my heroes in the field and they were arguing about my work."

In the end, Kurland fought a lonely battle. Ask most doctors and they'll tell you that the swine flu vaccine caused an epidemic of Guillain-Barré syndrome. Ask flu experts and they'll tell you that influenza vaccines in general cause the disease, albeit very infrequently, and that the swine flu vaccine was particularly adept at causing the illness. And ask any medical expert who worries about the dangers of a deadly flu outbreak and he'll tell you that the swine flu fiasco gives him pause.

"Among flu researchers, there are probably two events that have shaped how we think of the field," said Dr. Keiji Fukuda, who is the chief flu virologist at the Centers for Disease Control. "One is the 1918 pandemic—the sheer bigness of the thing. It is probably the largest number of people dying over that period of time from an infectious disease. The next comparable event would be to go back to the fourteenth century and the Black Death."

The second event, Fukuda said, is the swine flu episode. "It was sort of a counterpart to the 1918 epidemic," he said. "Something like 40 million people got the swine flu vaccine and several hundred developed Guillain-Barré syndrome. But no pandemic appeared. These things have haunted people."

The lesson from 1976, Fukuda concluded, "is that if a new virus gets identified or reappears, you don't want to jump the gun and assume a pandemic is happening."

# 7

# JOHN DALTON'S EYEBALLS

It was the customary Tuesday lab meeting at the Armed Forces Institute of Pathology. The setting was the usual pretty, but windowless, room in the bunkerlike building. The twenty or so scientists and technicians would file in, taking their places around a glass-topped wooden conference table while stragglers were consigned to chairs lined up along the walls. The agenda was the same as always—the person whose turn it was to lead the meeting would select a current scientific paper for discussion and present it to the group, sort of the scientists' analogue of a book club. That day, however, it was Dr. Jeffery Taubenberger's turn to lead the discussion and he knew it was going to be extraordinary.

Taubenberger, the boyish-faced leader of the group, fairly bounced as he walked into the room, carrying a small stack of photocopied scientific papers. One was on the histology of the eye, one on the biochemistry of color vision, and the third was the paper that had so intrigued him, an article that had just appeared in the February 17, 1995, issue of *Science* magazine, a leading journal. As the group settled in, munching sandwiches and sipping coffee or sodas, he began.

The story Taubenberger was going to tell was *Science* magazine's cover story. This one involved, of all things, John Dalton's eyeballs, and was to lead Taubenberger, in the most convoluted of ways, to the virus that caused the 1918 flu.

John Dalton, whose lean, somber, spectacled face stared out from the cover of *Science*, was a legendary chemist, born in 1766. It was he who proposed the atomic theory of matter—the notion that all matter is composed of invisible entities called atoms. He was so famous, in fact, that to this day an entire society, the John Dalton Society, exists to preserve his memory.

But Dalton had another legacy as well. He was color-blind. At first, he did not realize that he saw colors any differently than other people. He found out abruptly in 1794 when he mistakenly wore a brilliant red garment instead of the black called for by his Quaker faith. His fellow Quakers made him acutely aware of his mistake. Fascinated, Dalton delved into his own vision defect, wondering how and why he saw the world differently than others. He went on to describe, for the first time, inherited color blindness. Only his brother, he reported, seemed to see the way he did.

Dalton often said that, to him, grass and blood were the same color. Blue wildflowers were the same color as what others called "pink." It puzzled him—why could others see these colors called "red" or "green," "pink" and "blue," when he could not? Red, he wrote, appeared as "little more than a shade or defect of light." With such a famous sufferer, soon the very term for color blindness came to bear Dalton's name. "Daltonism."

Eventually, Dalton hit upon a possible reason why his world was monochrome. It must be that the fluid in his eyes, the so-called vitreous humor, was blue rather than clear, forcing him to see the world through a filter. Red and green would then appear to be the identical drab shade of gray. There was just one problem with the hypothesis: Dalton could think of no way to test it without removing his eyes and checking the fluid. Even he did not want to make that sacrifice, so he decided on the next-best solution. When he died, he said, he wanted his assistant to autopsy his eyes and examine them.

Shortly after Dalton's death on July 27, 1844, at age seventy-eight, his assistant, Joseph Ransome, did Dalton's bidding. He removed the eyeballs from the corpse and poured the vitreous humor from one eye into a watch glass. It was clear, "perfectly pellucid," Ransome wrote. Dalton's hypothesis was wrong. Then the assistant cut a nick in the other eyeball and looked through it to see if objects that were red and green appeared gray. They did not, leading Ransome to conclude that whatever had caused Dalton's color blindness must have happened outside his eyeball, in nerves connecting the eye to the brain. Because Dalton was so famous and his color blindness such a puzzle, his eyeballs were preserved in a jar and kept to this day by the John Dalton Society of Britain.

Now, Taubenberger told his group, more than 150 years after Dalton's death, the moment of truth arrived. Molecular biology had become so powerful that scientists, armed with a snippet of tissue from John Dalton's eyeballs, could decide why he was colorblind. Others had laid the groundwork, figuring out that color blindness is caused by a mutated gene, one that has gaps where genetic information is missing, preventing the gene from functioning. The question was, did John Dalton have that gene mutation? Was his color blindness extraordinary or pedestrian?

With a revolutionary technique known as PCR, for polymerase chain reaction, scientists could work with minuscule numbers of John Dalton's cells that they extracted from the interior of his eyeball, and decide if he did—or did not—have the mutated gene. That was what the *Science* paper, "The Chemistry of John Dalton's Colorblindness," was about. The answer, Taubenberger told his lab group, was that John Dalton had a perfectly ordinary inherited form of color blindness.

The more Taubenberger thought about the story of John Dalton's eyeballs, the more he thought that he, too, could do something like that. He had discussed it with a technician who worked for him, Ann Reid, that morning, at about six o'clock when they arrived at work. He dreamed that he could do an experiment that might make the cover of *Science* and that would delight lab groups around the world. His work might be the subject of hundreds, maybe thousands of journal club meetings and he might answer a question that had been lurking for decades. Yes, he was laboring in a little-known lab with a tiny group of scientists. And yes, he was unknown to the celebrities of science. But he was sitting on a molecular treasure trove. He could get preserved tissue from people who had died long ago—an entire warehouse of tissue. It was part of the Armed Forces Institute of Pathology, where he worked, and it had been established by President Lincoln himself. Lincoln had ordered that every time a military doctor examined tissue, from people who died or people with tumors or other diseases, the doctor had to send a piece of that tissue to the Armed Forces Institute of Pathology's repository.

Both military and civilian doctors provided tissue. "The pathology institute gets tens of thousands of cases submitted yearly for a second opinion," Taubenberger said. The patients tend to come from university research hospitals and, until recently, the military doctors provided their services free. Now they charge civilian doctors for pathology analyses, although they do not charge military doctors. But there is a catch, Taubenberger noted: "When they send in cases for an opinion, we get to keep representative samples for our archives. We keep everything—clinical histories, glass slides, paraffin blocks." Moreover, the tissue specimens preserved and stored decades ago tend to be as good as those preserved and stored today. Although medicine has been transformed since the nineteenth century, the art of preserving tissue specimens, developed more than a hundred years ago, has hardly changed. Even the so-called paraffin blocks remain. Primitive as it may seem, soaking a bit of tissue in formaldehyde and then embedding it in paraffin wax is still the method of choice for keeping specimens intact indefinitely.

Over the years, the armed forces' specimen collection swelled until it contained millions of scraps of tissue, preserved in thumbnail-sized chunks of paraffin wax or in jars of formaldehyde or smeared on microscope slides, all stored in cardboard boxes in a corrugated metal warehouse just a few miles from Taubenberger's lab. It was a veritable Library of Congress of the dead. Surely, Taubenberger thought, there must be some gems in there. The challenge was to think of the right question to pose.

Taubenberger wandered around the lab, asking himself whose tissue he should analyze. What tantalizing question could he answer? "We should do an analysis of somebody famous," he mused. "We must have the tumor of somebody famous here, President Garfield or someone."

He met that day with his boss, Dr. Timothy O'Leary, who was chairman of the department of cellular pathology, and with Dr. Marc Micozzi, the director of the National Museum of Health and Medicine at the Armed Forces Institute of Pathology, and told them his thoughts, inspiring the others to join him in trying to figure out what John Dalton-like project they should go after. Maybe they could dream up one having something to do with yellow fever, they thought, perhaps something involving the great turn-of-the century doctor, Walter Reed, who had figured out that yellow fever was caused by a virus transmitted by mosquitoes.

After all, the pathology institute was located on the campus of the Walter Reed Army Medical Center, which was run by the Army. But they could not decide what about yellow fever or Walter Reed himself would make a project. They thought of "camp fever" of the Civil War. Men had sickened and died like flies. Perhaps it would be possible to get some tissue from camp fever victims and ask if the illness really was typhoid, as medical historians have proposed.

Suddenly, the idea hit. "We were sitting around when someone thought of the 1918 flu," Taubenberger recalled. Each of the men realized that it was perfect. "That's it, that's the project," Taubenberger said. "It just resonated with us." They had the tissues from soldiers who died in 1918. More died of the flu than died in battle. Maybe they could find just the right tissue specimens, scraps of the lungs of young people who died in 1918 of the flu and whose lung tissue still contained fragments of the fatal virus. And maybe, if they used PCR, they could fish out those virus pieces and reconstruct the genetic code of the 1918 flu virus. They might learn the virus's identity. They might even learn why it was so deadly. They might solve the murder mystery.

Of course, it was a long shot. Even if they found people who had died of influenza in 1918, the virus in their lungs would be long dead and might have disintegrated before the lung tissue was stored. And even if the virus was still there, it was possible, if not likely, that too little of it had remained after storage for even the powerful methods of molecular biology to resurrect it.

The scientists told themselves to be realistic. "It started off as a low-key project. We thought the chances of success were pretty damn small," Taubenberger said. But it was hard for them to contain their enthusiasm once they began to plan the work. "We got suckered into it," Taubenberger recalled. "The more you look at it, the more interesting it is."

While they were waiting to begin, Taubenberger and Reid spent a couple of months reading about the 1918 flu. Both read

Alfred Crosby's *America's Forgotten Pandemic*. Taubenberger knew about the 1918 flu only vaguely, from his days in medical school. Reid had never heard of it. Both were stunned when they learned of the virus's devastation.

"I had no idea of the 1918 flu," Reid said. "But the more I read about it, the more committed I became to carrying on and sticking with it until we got some answers. I also was amazed as I talked to people outside. Anybody over the age of about sixty had a story to tell about the flu. My next-door neighbor's mother died of the flu when he was one year old. How could this huge thing happen and nobody ever talked about it?"

Even after the work began, Taubenberger kept quiet about the project, not mentioning it outside the confines of the pathology institute. He is by nature precise and wary of exuberant displays of confidence or enthusiasm. He prefers taking meticulous care with his work and biding his time.

Reid has a similar demeanor. Perhaps it is endemic among people who do the slow, tedious, precise work of molecular biology far from the competitive high-stakes biotechnology world. The molecular biology superstars can be publicity-hungry and voluble, but their job is to promote their work to other scientists and, often, to investors. People like Taubenberger and Reid were at the other end of the spectrum. They were not in the public eye and a substantial part of their job was to complete the tasks that had been assigned to them. While Taubenberger could and did pursue independent research, his main task was to set up a molecular diagnostics laboratory for the military.

In any event, even years afterward, even when they know how the project is turning out, Taubenberger and Reid shy away from revealing the extent of their emotions, preferring to focus on the details of the science and being sure to include cautions and caveats.

• • •

Taubenberger, as the lab leader, knew he was an unlikely person to go after the 1918 flu. But then again, he had always gone his own way, avoiding the obvious paths to scientific fame or fortune. He was one of those men who look like kids. Even in his late thirties, when he was head of a lab group, he could be mistaken for a young trainee. He has a round face, round brown eyes, and a shock of straight brown hair. He dresses like a student, in brown corduroy pants with an ID tag on a chain around his neck—no lab coat, no suit and tie for him. He drives an old Mercedes that he bought used in 1995 when his first child, a son, was born because he wanted a big, safe car. And he was completely unknown to the insular group of influenza researchers who had dominated the work on the 1918 flu.

Even Taubenberger's surroundings discouraged contact with the outside world. The Armed Forces Institute of Pathology is on the fringes of Washington. The Walter Reed campus, with its old brick buildings, is wedged in among blocks of small colonial houses on tiny lots, and nowhere near the scientific power centers. It is several miles from the crammed campus of the National Institutes of Health in the rolling hills of Bethesda, Maryland, and the towering white Navy hospital across the street where Presidents go for their physical exams.

Taubenberger's lab is in the most unusual building on the campus: a five-story gray concrete bunker that was built in the 1950s during the Cold War, designed to be bombproof. It has no windows and its walls are three feet thick. The idea was that when the atom bombs were on their way, the President and his cabinet would hunker down in the building, protected from the carnage outside. In fact, so taken were the military planners with the building that they proposed that all government buildings be made in the same way—the thick concrete walls, the lack of windows. The only thing that saved the federal government from an aesthetic fiasco was the discovery, just after Taubenberger's building was finished, that the structure would not protect anyone against hydrogen bombs. So Taubenberger's building remains one of a kind, a Cold War relic that has been turned into a warren of labs and offices.

The pathology institute is attached to a pathology museum, an all but deserted place that until recently was one of the last remaining specimens of old-time museums, without interactive exhibits or displays meant to capture people's fleeting attention. Instead, it held dusty jars with such things as a hair ball, taken from the intestines of an eleven-year-old girl who had been chewing on her hair all her life. It had terribly deformed fetuses in bottles. It had an entire leg from an elephantiasis victim in a large jar. It had a display of iron lungs from the polio era. But what it did not have was visitors. While it once had been located on the Mall downtown, with the other major federal museums, it was evicted more than twenty years ago to make room for the Hirshhorn modern art museum. Most tourists, swarming over the grassy downtown Mall and into the stately museums that line it, have never heard of the National Museum of Health and Medicine, and even if they had, few would venture so far afield to see it. Although the pathology museum has had an update and remodeling meant to make it more modern, with an exhibit on AIDS and the ubiquitous interactive displays, its isolation has determined its fate.

Anyone who journeys to the building to visit Taubenberger must check in with a security guard who sits behind a gray metal desk in a drab hallway. Taubenberger himself—he has no secretary—comes down to escort visitors to his tiny office on the third floor, with its faded and stained blue rug and scuffed furniture that Taubenberger salvaged from the loading dock outside where discarded desks and tables and chairs accumulate. Every flat surface is piled high with papers and articles from scientific journals. And every vertical surface is dotted with yellow Post-it slips on which Taubenberger has scrawled notes and reminders to himself.

• • •

Taubenberger's training was as unconventional as his position. Ordinarily, young scientists scramble to get on the fast track. The idea is to go to the best university that will admit you, then you travel to the best graduate school you can get into. From there you do a postdoctoral fellowship in a prestigious lab, if you can. Then you travel again, getting a junior faculty position somewhere and you work like crazy to do something spectacular, so good that you will get a tenured position, perhaps at another university. In the meantime, you try to publish, publish, publish. You travel to meetings and give talks on your work. You latch on to scientific superstars and hope to get noticed. It is a peripatetic life, one with no guarantees of success. And it is a life that is completely foreign to Taubenberger. He traveled an alternate route.

Taubenberger was born in 1961, in Germany, the third son of a career Army officer. His father had helped build one of the first transistor computers, a "mobile" computer that took up five semitrailers. For its time, around 1956, it was a revolutionary machine. "My father was already a computer person in the vacuum tube age," Taubenberger said.

The Taubenberger family lived in Europe and California, eventually settling in Fairfax County in northern Virginia, part of the sprawling and bland Washington suburbs, after Taubenberger's father was posted at the Pentagon. Ever since he was four or five years old, Jeffery Taubenberger knew he wanted to be a scientist, the only question was what area of science. He liked nuclear physics, then chemistry, and finally settled on biology. "Life is so complicated—it's really cool," he said, explaining his choice in his usual way, devoid of introspection.

But the Fairfax public schools did little to encourage Taubenberger, the childish-looking science lover whose other passions—classical music and composing—also were not part of the high school culture. The pace of the courses was slow, the work was just too easy, and Taubenberger, complaining that the school "wasn't very academically rigorous," got little sympathy from his teachers or the students who were listening to rock music and cheering the football team.

There was, however, a way out. Taubenberger discovered that he could escape from his high school, skipping his senior year, and actually start college by enrolling in a program at George Mason University in Fairfax. It was near his home, so he would not have to worry about moving away or paying for a dorm room. And it promised Taubenberger a chance to take classes that might actually challenge him. So at age fifteen he began classes as a college freshman. That summer, he got a job at the National Institutes of Health, working on mouse mammary tumor viruses, a type of virus that could cause breast cancer in mice. It was 1977, and in those days researchers thought that viruses might hold the key to human cancer (they turned out to play no discernible role in common human cancers, like breast cancer, although they contribute to several cancers, such as cervical cancer and Kaposi's sarcoma). At the time, the brightest biologists were studying tumor viruses and Taubenberger was being introduced to the cutting edge of one of the hottest fields of biology. He was hooked and continued working in the lab part-time after school began again in the fall.

"It was a great experience," Taubenberger said. He thought of transferring from George Mason and applying to a prestigious college, Harvard or Princeton, perhaps. Then he decided that he'd rather stay where he was so he could keep working on those tumor viruses. That experience was worth at least as much as an education from one of those better-known schools, he reasoned. "I decided to stay at home and work at the NIH," he said matter-of-factly. He had a career plan. What he wanted to do was to get a Ph.D. in biology and then eventually head a lab, like Dr. William Drohan, the man he worked for at the National Institutes of Health. Drohan explained to him that credentials matter in the scientific world and that one credential, an M.D. degree, can open doors. Biologists who had M.D. degrees as well as Ph.D.s seemed more impressive, it was easier for them to get grants, and they were offered better jobs. "I had never thought of being a physician," Taubenberger said. Drohan, he adds, "convinced me."

As soon as he graduated from college, Taubenberger went to the Medical College of Virginia in Richmond, a state school, with a tuition that was much lower than a more prestigious private college—a large part of its attraction for the frugal young man. His course of study was an M.D.-Ph.D. program in which students earned both degrees simultaneously. While many medical students complained of the grueling classes, the ever-ebullient Taubenberger was entranced. "I really got enamored of medicine," he said. "It's intellectually stimulating. It's fun."

The Ph.D. part of the program had its own requirements. As is customary in Ph.D. programs in science everywhere, the program made it obligatory that everyone who gets a degree has to discover something new, complete a research project that displays intellectual rigor, and end up with a nugget of information that no one had found before. Each Ph.D. student has to work with an advisor, who serves as a mentor and guide and who, if all goes well, helps the student learn how to do original work and how to complete it well enough so that a Ph.D. committee would deem the student worthy of the degree. Taubenberger set out to find the perfect advisor, the person at the medical school whose work was fascinating and who could help him pose an answerable question and help him learn the techniques he would need to answer it.

"I spent the first six months of medical school going around and talking to everyone," Taubenberger said. He settled on Dr. Jack Haar, a professor in the anatomy department, whose interest was in how T cells, which are a class of white blood cells, develop in a spongy organ called the thymus. "I really liked him," Taubenberger said of Haar. "I went to his lab and worked there on the side and during my summer breaks."

The usual sequence for M.D.-Ph.D. students was to spend the first two years taking classes with the medical students, then do Ph.D. work, then finish medical school by doing clinical work. Taubenberger rearranged the sequence, doing his clinical work right after the two years of medical school classes and then rushing to the lab where he spent all his time doing his Ph.D. research. He got an M.D. degree in 1986 and a Ph.D. in 1987 for work on how bone marrow cells migrate to the thymus to become T cells. Not only did he do the Ph.D. work, but he got awards and scholarships along the way, including the school's first prize for medical research and an award for being an outstanding cell biology student.

With his degrees in hand, qualified in just the way his mentor had suggested, Taubenberger had to decide on a career. "I didn't know what I wanted to do," he said. "I had become very interested in pediatrics and I thought I would do pediatric hematology. But things were going very well in the lab, so I stayed one more year, until 1988."

At that point, he began to worry about being a doctor again. "I thought I would do a residency in pediatrics, but I had been in the lab for four years without ever seeing a patient." He envisioned the worst-case scenarios, the patients who might suffer if he made a mistake. "Being an intern in the newborn intensive-care unit would be really scary," he said. Inevitably, he seized upon pathology, where he at least would not have to be examining live patients and making life-or-death treatment decisions. Pathology, he decided, was "kind of an intermediate thing where I would still be involved in medicine but more research-based." He began looking for places where he could do a residency and discovered immediately that there was a pathology program at the National

Cancer Institute, part of the National Institutes of Health. It accepted only three people a year. Taubenberger applied.

He set his heart on the fellowship, despite the long odds against him. After all, it was the National Institutes of Health, the place where he had gotten his start. "Obviously, that was a great place to do research. That was home."

Shortly after Taubenberger had his interview for the position, he got the good news: he was accepted into the program. He began in 1988, with residency training at the National Cancer Institute; he finished in 1991. Afterward he managed to stay around, doing a postdoctoral fellowship. When that ended, his advisor, Dr. Ada Kruisbeek, a renowned Dutch immunologist, gave Taubenberger a chance to start on the usual scientific fast track. She told him that she was returning to Holland, where she had been offered a prestigious position, and asked Taubenberger to go with her as a postdoctoral fellow. After considering the offer, realizing that it could be the start of a life in the high-stakes game of molecular biology, Taubenberger thought better of it.

"I had just gotten married. My family is here, my wife's family is here," Taubenberger said. He decided to stay where he was. He managed to get another temporary research position at the National Institutes of Health. Finally, in 1993, he was offered a more permanent job that let him stay in Washington: he and Dr. Jack Lichy, who also was working at the NIH, got jobs setting up a molecular pathology lab at the Armed Forces Institute of Pathology, using the tools of molecular biology to make diagnoses for pathologists. Six months later, they had the lab up and running. At the same time, Taubenberger had a research lab where he could pursue questions in basic science.

A year after he began, in early 1994, Taubenberger was promoted to chief of the Division of Molecular Pathology, a job whose title suggested something grander than the reality, supervising a group of twenty people. "It's a very small shop," Taubenberger said. He carved out a satisfying life for himself.

Taubenberger leaves for work each day at 5:50 in the morning, partly to avoid the abominable Washington rush-hour traffic, driving his car from his home in Virginia to his lab. He spends his spare time with his wife and young children, whose photos he uses as a screen saver for his computer, so they are always on his mind. In his spare time, he composes music, his hobby. He wrote an opera when he was in college. "The overture got performed," he said wryly. He wrote a symphony when he was in medical school, which the Richmond Community Orchestra performed while Taubenberger played with the orchestra, on the oboe. He wrote a piece for a woodwind quartet and, as a gift for his wife, he wrote a piece for a string quartet that was played at his wedding. He's working on another string quartet in honor of the birth of his son. But he composes only as a labor of love. "I still have no luck getting things performed," Taubenberger said.

One day, in late 1993, Dr. Thomas Lipscomb, a lieutenant colonel in the Army who is a veterinary pathologist, came to visit. He had a problem and hoped Taubenberger and Lichy could provide a solution. Over the past decade, he explained, there had been strange outbreaks of deadly illnesses in marine mammals. In 1987, bottle-nosed dolphins living along the coast of New Jersey had been dying. The disease—or whatever it was—began moving south, killing dolphins off the coast of St. Petersburg, Florida, in March 1988. Half of the inshore Atlantic U.S. dolphins died, some ten thousand animals.

The Environmental Protection Agency started an investigation and issued an initial report saying that the deaths were caused by a red tide, an infestation of tiny dinoflagellates that turn water a rusty hue and that release a toxin. Red tides are a sign of pollution and the hypothesis was that pollutants were ultimately killing the dolphins.

Lipscomb did not like that explanation, citing two reasons why he thought it could not be true. First, he told Taubenberger, red tides affect a slew of animals all at once. This epidemic was killing only dolphins. Second, red tides kill smaller animals—their toxin has never been known to kill an animal as large as a dolphin.

Instead of a red tide, Lipscomb thought that a virus was killing the dolphins. After examining tissue from hundreds of dying or dead dolphins that had washed ashore, he saw a clear pattern: the animals' brains, lungs, and lymphoid tissues appeared to him to be infected with a morbillivirus, a type of virus that causes measles in people and distemper in dogs.

Embellishing his hypothesis, Lipscomb pointed out that there were other strange incidents in which marine mammals were dying of what looked like a virus. In 1988, seals in a lake in Siberia sickened and perished. In 1990, harbor seals along the northern European coast succumbed. In 1991, striped dolphins in the Mediterranean died and European scientists isolated distemper-like viruses from the animals. In 1993, dolphins living in the Gulf of Mexico died.

Now Lipscomb, seeking the molecular evidence, came to Taubenberger with a challenge. He had decayed dolphin tissue, some so putrefied it was not even possible to examine it under the microscope, tissue that was "really gross," Taubenberger said. Could Taubenberger and the Molecular Pathology Division use PCR to fish out a morbillivirus, if it was present?

"I thought the chances of getting a viral RNA out of such tissue was zero," Taubenberger said. But he did not refuse Lipscomb. Instead, he asked Dr. Amy Krafft, a young molecular biologist who had just joined the group, to work on the dolphin project. The methods available were precise, but she had to hone them, pushing them to their limit. She worked and reworked every step of the intricate biochemical experiments that might allow her to retrieve the viral RNA, optimizing each step until she had a perfectly efficient system. And then she did the impossible—she retrieved the virus's RNA, showing that Lipscomb was correct. In the end, she isolated a new species of morbillivirus, demonstrating that they, and not a red tide, had killed the dolphins.

That experience was fresh in Taubenberger's mind when he thought about going after the 1918 flu. It would be similar work. The flu virus, like morbilliviruses, used the genetic material RNA. Flu viruses were about the same size as morbilliviruses, containing about 15,000 bases—the chemicals adenine, guanine, cytosine, and uracil that link together in long chains to make RNA. And the morbilliviruses share many features with flu viruses. That meant that since Amy Krafft had been able to get a virus out of the horribly decayed dolphin tissue, she might be able to obtain a flu virus out of the tiny slivers of lung tissue from the tissue repository. Now the scientist had to find out whether the warehouse had any tissue from people who died of the 1918 flu.

The records of all tissue specimens stored since 1917 were available on a computer, a total of 3 million specimens. But it was entirely possible that there was no lung tissue from soldiers who died of the flu in 1918. The epidemic struck during wartime. "Lots of people who were called in to do autopsies weren't trained in pathology," Taubenberger said, making the search problematic from the start. Why would a military doctor take the time and trouble under such chaotic conditions to carefully cut out a piece of a dead man's lung and send it off for storage?

Nonetheless, it was worth a try. The lab group would figure out how to search for the most likely flu victims and then send in its request. Taubenberger decided to look for people who had died within days of becoming ill, reasoning that if they looked simply at the lungs of people who had died of the flu, they might get people who had started out with the flu and then gotten a bacterial pneumonia that ultimately killed them. By the time the flu victims died, possibly a week or more after they were infected with the flu virus, the virus might be long gone from their lungs. Only bacteria would remain.

Taubenberger put in a request. Two days later he got a computer printout describing what was in the warehouse, indicating that there were seventy specimens from people who had died of the 1918 flu. The specimens were tiny bits of lung tissue, soaked in formaldehyde and encased in little blocks of paraffin wax. Along with the specimens were the flu victims' medical records, which stated when the people got ill, how and why they died. Taubenberger eagerly pored over the documents. Six of the victims met his specification—that the deaths occurred within days of the illness. The scientists were amazed and overjoyed.

"It was like *Raiders of the Lost Ark*," Taubenberger said, referring to the Steven Spielberg film. "We found the ark of the covenant."

Here was a warehouse with millions of specimens from long-dead patients tucked away in cardboard boxes, filed and forgotten. Somewhere in the vast storage area was the 1918 flu virus, hiding out in a thumbnail-sized scrap of lung tissue. No one had thought to look for it for nearly eighty years. And even if they had tried to search for it, the virus was so well hidden in the tissue that no one would have been able to find it. But now, acting on a spark of inspiration from a paper on John Dalton's eyeballs and hoping to use methods that the Molecular Pathology Division had developed since 1989 and that Amy Krafft had refined to look at decayed tissue from dead dolphins, a group of scientists that no one had ever heard of, investigators who had never before done research on influenza, had found the first hot clue to the murderer of 1918.

Once they knew there was at least a chance that they could find the virus—at least there were specimens from the lungs of victims of the 1918 flu—the scientists could hardly wait to start. But,

Taubenberger said ruefully, "you have to go through some government red tape. You can't willy-nilly do anything." He had to write a research proposal, and since he wanted to use human tissue, he had to get special approval from a committee that rules on all uses of human tissue. It took a couple of months—time he and Ann Reid spent learning as much as they could about the 1918 flu. The moment they got permission to begin, they ordered the tissue, oddly enough never visiting the warehouse to see the precious repository for themselves.

The warehouse is a few miles from the pathology institute, just over the line dividing the District of Columbia from Maryland. It is a low building of corrugated steel, fireproof, of course, lined inside with rows of metal shelves. The shelves contain the specimens, row after row of cardboard boxes containing paraffin blocks with tissue inside, or jars of formaldehyde-preserved organs or microscope slides of artfully stained cells. It is run by Al Riddick, a genial, but muscular man who generously shows a visitor how he retrieves the specimens. When Riddick got the call from the pathology institute for the lung tissues from victims of the 1918 flu, he matter-of-factly climbed a ladder and took down cardboard boxes from the metal shelves. Inside each box was a chunk of paraffin wax, about the size of a fingernail on its end and about a quarter inch thick, and encased in it was a sliver of lung tissue from a soldier who had died in 1918 of the flu.

Luckily, Reid thought, she knew how to work with tissue encased in paraffin. That required meticulous care, and it would stand her in good stead for the flu project. There was so little material—and such a small chance of finding anything—that a mistake that destroyed any lung tissue could be disastrous. Reid also knew that many lab directors would have been nervous placing her, with her minimal experience, in charge of such a task.

When Reid had come to the lab, in 1989, she had been a very low-level technician, with only a year and a half of experience. "I had never really worked independently at all. I didn't have any agenda other than that I would do what I had been told to do," she recalled. But she gradually took on more and more responsibility as she became an expert at working with scraps of tissue encased in paraffin. She had, for example, been asked to check tissue from people who had had lymphomas, cancers of the lymph nodes, to see if she could extract Epstein-Barr virus from their cells. And she had been given other tissue, embedded in paraffin, from people who had had other types of cancer and asked if she could extract the cells from their waxy case and do a genetic analysis, looking to see if the cancer cells' chromosomes had broken and rejoined, which is a way that some cancers are instigated. By 1995, Reid had probably had as much experience as anyone in the world on doing PCR from paraffin-embedded tissue. Her experience was ideal for this project. At the very least, she would be able to start the process of looking for the 1918 flu virus. But as she was well aware, it was not at all clear that she, or anyone else, could complete it. They would be looking for what might be literally a single molecule of flu virus. It would be a test that would push the limits of molecular biology.

With some trepidation, once they got permission to begin to study the tissue from 1918 flu victims, Taubenberger and Reid set to work. Reid noted the date in her lab notebook: March 19, 1995.

The scientists began by slicing off a few exquisitely thin slices from the paraffin block. Using a sharply honed razor blade, working by hand, they shaved off pieces that were the thickness of a single cell, so fine they were much thinner than paper. The slices contained a total of no more than about 2,000 cells. Within perhaps one of those cells might lie the shattered remains of a flu virus.

With the tissue slices in hand, Reid's first task was to dissolve the wax away from the tissue. To do that, she put the slices into a small test tube and added xylene, a solvent that dissolved the wax, leaving the tissue floating free.

Then she had to separate the xylene from the lung cells, so she spun test tubes containing the mixture in a centrifuge; the cells clumped at the bottom of the tubes. After pouring the organic chemical out of the test tubes, Reid washed the cells with alcohol, to draw out any xylene that still clung to them.

The next step was to separate the cellular debris—membranes and proteins—from the genes and gene fragments. Most of the genes were from the cells of the dead man, but perhaps, if they were lucky, there also would be genes from the virus that killed him. To isolate the genes from the mixture, Reid put the clump of lung tissue into a test tube and added a salt solution and an enzyme called proteinase K that chewed up proteins. She added a little detergent, to dissolve the fatty cell membranes. About four hours later, she came back to the test tube and did an organic extraction, the lab equivalent of making a salad dressing.

Because of their chemical composition, genes dissolve in water, whereas the protein fragments and bits of fat from the cell membranes dissolve in oil. To separate the genes from the cell membranes and proteins, Reid added an oily mixture to the solution—chloroform and phenol. She shook it up so that the protein fragments and fats in the mixture of molecules from the lung tissue would dissolve in the oily layer and the fragments of genes would stay in the watery layer. Then she let the mixture settle so that the oily layer floated on top of the water. She poured off the oil. Next she centrifuged the mixture and removed the watery solution off the top and was left with a thin liquid that contained gene fragments from the lung tissue, from any bacteria that might have been in the lungs, and from the 1918 flu virus—if it was present.

To separate the pieces of genes from the watery mixture they were dissolved in, Reid added alcohol to the solution. She placed the test tubes in a centrifuge and set the machine spinning so that the shards of genes were forced to the bottom of their containers, forming a tiny pellet. She poured the liquid out of the tubes and added salt water to dissolve the pellets at the bottom, leaving her with genes in a salt solution that lent itself to genetic analysis.

After days of work, Reid was ready to start to see what sorts of genes might be in the lung tissue. "Someone in the lab said that molecular biology is moving tiny droplets of clear fluids from one tiny test tube to another all day long," Taubenberger remarked. "It requires a lot of faith since you never can see what you are doing."

It was time to use PCR, the miraculous method that can make millions of copies of a single fragment of a gene floating in a solution. Taubenberger and Reid decided to look for a piece of a flu virus's genetic material called the matrix gene, choosing it because it changes very little, whereas other viral genes are constantly being altered by mutations. The matrix gene directs cells to make a protein that the virus uses as a scaffolding, a firm structure propping up the virus's soft, fatty membrane coat.

Their idea was to use a tiny bit of a matrix gene as a sort of fishhook—a way to snare the 1918 flu virus. With the gene captured, they could use PCR to start copying a piece of it. But if pieces of the matrix gene were to be the hooks for the 1918 virus, Reid and Taubenberger needed to find the best pieces, those that were least likely to vary from strain to strain of flu virus. To select such gene segments, they lined up published matrix gene sequences from a variety of influenza viruses, picking out regions that were virtually identical. From those constant regions they designed their fishhooks, or primers.

With PCR so indispensable to molecular biologists, commercial companies have sprung up to custom-make primers for scientists. Reid and Taubenberger used Integrated DNA Technologies of Coralville, Iowa, sending the company a fax with the exact genetic sequences of the primers they wanted. Within days, they received them by Federal Express, flecks of dried white powder in the bottom of small test tubes. All that Reid had to do was add sterile water and she had her fishhooks.

Reid started her PCR experiments by mixing the custom-made primers with genetic material from about a half dozen victims of the 1918 flu. If the primers found any matching gene fragments in the tissue, an enzyme would make millions of copies of the gene segments. Reid could spot these copies by labeling them with a radioactive probe as they were being made. Those minuscule pieces of genes, far smaller than anything that could be seen with a microscope, would burn black marks onto a piece of X-ray film. The question was: Would there be any radioactive gene fragments in the test tubes? Would the primers have found any shards of the 1918 flu?

At the end of the day, Reid put a 15-by-17-inch strip of X-ray film against the gene fragments and went home, waiting while the fragments, if they were there at all, seared black bands onto the film. The next day, barely allowing herself to hope, she went into the tiny darkroom and pulled out the film. Taking it into her lab, she held it up to the fluorescent light.

The experiment failed. Where she had hoped to see gene fragments from the 1918 virus, there was nothing. The film was blank. Reid knew it was not a failure of her laboratory technique: she had done a parallel experiment with genes from a known virus, called PR34, for Puerto Rico, 1934, the place and year it was isolated. It was the oldest flu virus whose sequence was known, leading Reid and Taubenberger to hope that its genetic sequence might be close to that of the 1918 flu. If the primers could find pieces of the PR34 matrix gene, they might be able to find fragments of the 1918 flu. That experiment worked. She saw the matrix gene from the 1934 virus.

Reid tried the tedious experiment again. And again. She was using tissue from a dozen of the samples from the warehouse, scraping off precious and irreplaceable lung cells from the paraffin blocks. But there was no sign of the 1918 virus. The problem was that she did not know what was wrong. Maybe there was no flu virus in the tissue. Or maybe her methods were not good enough.

"It was terribly discouraging," Reid said. "I spent a lot of time in prayer over this. At a certain point, you've tried everything and you don't have any more ideas."

In June, after more than a year of failure, after seeing nothing on the X-ray films day after frustrating day, Reid and Taubenberger decided they had to try something different. It was time to back up and try a simpler experiment, one that would help them know if their task was even possible. They would see if they could get flu virus out of preserved lung tissue that came from more recent flu victims. They would choose a flu strain whose genetic sequence was known, so if they did find something in the lung tissue, they would know whether it was the flu virus and whether the genetic sequence they determined by their methods was correct. It would be a detour in their path to find the 1918 virus, but now, clearly, it was necessary.

The next question was, Which flu virus should they look for? They settled on the one from 1957 which had swept the world, killing about 60,000 Americans, although it was nowhere near as deadly as the 1918 flu. The virus's genes were known, which made it a good test case. And the armed forces pathology warehouse should have lung specimens from that flu's victims. The samples would be forty years old—a good test to see if viral genes withstand the rigors of decades of storage. If they do—if Reid and Taubenberger could fish flu genes out of those samples—they could have renewed hope that the viral genes from 1918 could be found in the older material.

The warehouse faithfully delivered paraffin blocks with lung tissue from victims of the 1957 flu. To begin the analysis, Taubenberger and Reid called in Amy Krafft, who had become so expert at getting genetic material out of decayed dolphin tissue. She began working with the tissue from the 1957 flu victims, using all she had learned, skillfully pushing the limits of the extraction technique. At the same time, she prepared solutions from a half dozen victims of the 1918 flu. Krafft gave all of the solutions to Reid for PCR.

Then Reid took over. She used the fishhooks for the matrix gene again. The next morning, at 6:30, Reid went into her darkroom and pulled out the X-ray film. She took it into her lab and held the film up to the light. This time she saw something—a dark band that could only have come from a matrix gene from one of the samples that Amy Krafft had prepared. After looking at blank films every morning for fifteen months, she was incredulous, and almost afraid to hope that what she was seeing was real.

Reid ran to Taubenberger's office, X-ray film in hand. "We went absolutely nuts," she said. "We thought we had it." They told each other to calm down, that it can't be real. It probably was a contaminant, a stray bit of gene that got in the way. PCR is so sensitive that all it takes is one molecule of yesterday's reaction floating into today's experiment and an error can be made. But they could see that whatever they had found, it was not the same as the matrix gene fragments from PR34.

Reid set to work, finding the sequence of the mysterious gene fragment. It was just 70 bases long, about a twentieth of the size of a flu gene. Now Reid had to do yet another experiment, adding those tiny fragments of genes to little circular viruslike strings of genes, called plasmids, that can infect bacteria. The bacteria would grow and multiply, replicating the plasmids every time they divided. Reid could then isolate the plasmids, use enzymes that act like molecular scissors to cut out the genes she was interested in, and, from there, decipher the sequence of the genes.

When she had the sequence, she was ready to see what flu virus it came from. She went to her computer and called up the

National Library of Medicine's Web site and navigated to a program there called "BLAST" that will compare any gene sequence to all known gene sequences and report back on what are the closest matches. Reid typed in the sequence of her gene fragment. The answer came back immediately: it was a perfect match for the matrix gene of the 1957 influenza virus.

"Initially, we were really bummed," Reid said. "We thought we were sequencing a 1918 case, but it turned out to be from 1957." But, she said, she and Taubenberger soon realized what they had done and were chilled at the thought.

"It was wonderful," Reid said. "In a way it was more wonderful than getting the 1918 flu because now we knew it was possible. If the genes had lasted forty years, there was no reason why they wouldn't last eighty."

The work went quickly then. Reid, Krafft, and Taubenberger returned to their paraffin blocks with lung tissue from victims of the 1918 flu. This time, they concentrated on the lung tissue of Roscoe Vaughan.

Vaughan was the twenty-one-year-old soldier who had died in Camp Jackson, South Carolina, in September 1918. When they looked at his medical record, "we knew it was the flu," Taubenberger said. "He had a rapid onset of illness, a high fever, chest pain, cough." And, most important, he had died suddenly.

Vaughan had become ill and reported to sick call on September 19. He died at 6:30 in the morning of September 26. And by 2 p.m. of that day, Captain K. P. Hegeforth had done an autopsy, noting that Private Vaughan's lungs contained about 1¼ cups of clear fluid in his chest cavity and that blood had been seeping out onto the surface of his left lung. He noted that the air sacs in Vaughan's lungs were filled with fluid; the young man had literally drowned in his body fluids. Finally, Captain Hegeforth removed a scrap of tissue from Vaughan's lung and carefully preserved it in formaldehyde and paraffin. He sent it to Washington for safekeeping, where it remained until Taubenberger called it up eight decades later.

Taubenberger peered at the tissue samples from each of Vaughan's lungs. "This was one of the few cases that struck me as being a likely candidate for being positive" for the 1918 flu virus, he surmised. "I was looking for cases in which the clinical course leading to death was less than a week. A lot of the material was from people who died of bacterial pneumonia. I didn't think those lungs would still harbor the virus. In this case, there was a very unusual pathology; the left and right lungs were very different. This was noted clinically on his chart and was also noted at autopsy." Vaughan died of massive pneumonia in his left lung, Taubenberger explains, noting that people can go into respiratory failure and die if one lung is suddenly and catastrophically rendered nonfunctional, whereas if one lung gradually fails, the body can adapt. Vaughan's right lung was different. "It had very, very subtle focal changes, evidence of very early inflammation that was consistent with a primary influenza infection," Taubenberger said.

Carefully, Reid and Krafft began, using a slice of Vaughan's preserved right lung. Once again, Krafft separated the genes from the rest of the lung tissue and debris and then Reid searched for genes from the flu virus, using a matrix gene as the fishhook.

Finally, the moment of truth had arrived. The fragments of the flu virus gene would have radioactive labels that would allow Reid to see if they were present in the salty fluid mix of plasmid genes. She separated the gene fragments on a thin gel by running an electric current across the gel—the gene fragments spread out according to their size and chemical composition. Then she put a piece of photographic film over the gel—the radioactive labels would create black spots on the film.

Reid hung the film up against a light box and stared at it. And she saw them—the black marks that showed that the matrix gene from the 1918 virus gene was there. She felt a chill run down her spine. She now knew her snare could catch the 1918 flu, the killer she had been stalking. "Very few in science are given that kind of moment," she said.

Reid began searching for one viral gene after another, using snares to see what other genetic footprints of the 1918 killer were hiding in Vaughan's lung tissue. Everything worked—she began picking up traces of every viral gene she looked for. The neuraminidase, the nucleoprotein. The matrix genes, M1 and M2. Soon it was August and Reid felt she had to go on a long-planned monthlong vacation with her family, to a cabin in Vermont. She had labored over the project for a year and a half and, at least, she was leaving with the knowledge that the method could work: the virus could be snared.

Taubenberger took up the project in early August 1996, going after what many felt was the key gene, the hemagglutinin. Reid had no telephone in the cabin where she was vacationing, but as he got results, Taubenberger kept calling her mother-in-law and father-in-law, who lived nearby, asking them to convey messages to her.

Reid was elated, and also stunned by the success of the project. She knew that most scientists would not even acknowledge her as a colleague, yet here she was, a key part of the team that was hot on the tracks of the elusive killer virus. After all, she was just a technician, she did not have a Ph.D. "It's a very hierarchical field," she said. "I have had experiences before this where someone will be talking to me and will say, Where did you postdoc? I tell them that I did not postdoc and they would turn around and walk away."

Nonetheless it was Reid, perfectly prepared by her work with the paraffin-encased tissue, who was able to get the flu virus fragments out of the lung tissue. The others in the group seemed just as fortuitously ready. There was Amy Krafft, prepared by her work with the dolphins. There was another lab scientist, Dr. Thomas Fanning, who just happened to have worked on genetic relationships between species and whose expertise would be needed to figure out the etiology of the 1918 flu. There was Dr. Timothy O'Leary, their boss, obsessed with the idea of making use of the tissue repository. And finally, there was Taubenberger, the molecular pathologist whose imagination was sparked by the article on John Dalton and who then had the vision and persistence to search for the 1918 virus.

"It's really amazing that we had all the right people here to make it work," Reid said. "Looking back, it was like it was meant to be."

By October 1996, the group was ready to tell the world that they had genetic evidence of the 1918 flu in the preserved lung tissue of Roscoe Vaughan. They would write a scientific paper explaining their astonishing result. They did not have the entire genetic sequence of the virus yet, but they had shown that they could pull the viral genes out. That meant that they eventually might know the virus in its every detail. They had the tools to unmask the murderer and even find its deadly weapon.

The scientists decided to send their paper to *Nature*. Certain that what they had found was electrifying, they gave the journal editors advance notice that their paper was on its way. "I E-mailed the Washington office of *Nature*," Taubenberger said. "An hour and a half later, I got a phone call from London," from the editorial offices of the journal, "saying, 'This is really great, send it right in.'" He did, expecting that "this paper would get sucked right in," published posthaste. But, to Taubenberger's amazement, upon receiving his paper, *Nature* sent it right back, rejecting it without even mailing it to experts for review. The journal included its standard rejection note, explaining that the flu paper was not interesting enough for review.

Baffled, Taubenberger then sent the paper to *Science*, *Nature*'s chief rival and an equally prestigious journal.

But *Science* magazine, apparently, was just as aloof. "We sent the paper to *Science* and they just bumped it," Taubenberger said. Why? Perhaps the scientists looking at the paper questioned the group from the Armed Forces Institute of Pathology. "It gave the flu community a shock to think that a non-flu person was working on this flu project," Taubenberger speculates. "In the flu community, people may not have heard of us." Only after some senior scientists intervened on Taubenberger's behalf was his paper sent out for review. Then, he said, the reviewers were enthusiastic about the paper and it was accepted for publication. But Taubenberger was shaken by the experience. "It scared the hell out of us," he said. "I thought it would never get published." After all, he adds, he had no experience with high-profile science and he just assumed that if he did something really important, major journals would jump at the chance to publish it.

"It was a very odd experience," Taubenberger reflected. "We had always been writing papers that were run-of-the-mill. Nothing splashy. This was the first time I ever wrote an eye-opening paper."

Taubenberger had come across the surprising capriciousness of major journals, which have been known to reject out of hand revolutionary papers while routinely publishing the mundane. It happened to Dr. Ryuzo Yanagimachi, a scientist at the University of Hawaii in the summer of 1998, when he managed to clone mice and even to clone the clones. This was at a time when some leading scientists were harrumphing loudly about Dolly, the first animal that was cloned from the cells of an adult sheep. How did they know that Dolly was really a clone? the critics asked. Perhaps there was a mix-up of cells in the lab. And when would the next clone appear? A single example is an anecdote, not an experiment, the critics said.

Yanagimachi managed to do the impossible, to make cloning seem easy, creating dozens of clones. He sent his paper to *Science*, reasoning that it was a choice between *Science* and *Nature*, the two leading journals, and he, as an American, wanted *Science*, since it was an American journal. Like Taubenberger, he expected that the journal's editors would rush his paper into print. Like Taubenberger, Yanagimachi was stunned by the journal's response. *Science* returned his paper without sending it out for review, and included a note saying the work was not of general interest. Then Yanagimachi sent his paper to *Nature*, where it hung around for months, with reviewer after reviewer looking it over before, at long last, it was published, to worldwide acclaim.

But Taubenberger was unaware of the games journals and reviewers—who sometimes are ignorant, sometimes are jealous, sometimes have undisclosed conflicts of interest—can play. And when his paper was treated so cavalierly, he was stung.

Ever cautious, Taubenberger waited until his paper finally was accepted and on its way to publication before contacting Alfred Crosby, the historian whose book had inspired Reid and him in their lonely and frustrating year and a half of effort. Going to the University of Texas Web page, Taubenberger searched an E-mail address for the history professor and sent him a note. Taubenberger reminded Crosby that in his book he had said that unless there is a time capsule buried somewhere, the 1918 flu virus is gone forever. Taubenberger told him that he had found that time capsule—it was a piece of lung tissue from 1918 that was in the armed forces pathology warehouse.

Taubenberger's *Science* paper on the 1918 flu finally appeared, in March 1997, catapulting the unprepossessing researcher, who had spent his life in obscurity, into a media spotlight. "The phone was ringing off the hook. All of a sudden, eighty people wanted to interview me. I was live on national TV, I was on National Public Radio. It all happened so fast, it was really odd. It was completely crazy."

But that was just the beginning of the oddness. Within two months, a mysterious death in Hong Kong would make scientists quiver with dread. All of a sudden, it looked as if another deadly flu epidemic might be emerging.

# 8

# AN INCIDENT IN HONG KONG

Dr. Nancy Cox was on vacation in Wyoming when she got the call from her lab in Atlanta. Virologists there had done what they thought would be a routine test to determine the strain of an influenza virus isolated from a patient that past May. The sample had been stored at the lab for about a month and taken up in its turn for analysis. But when Cox, who directs the influenza lab at the Centers for Disease Control and Prevention heard the result, her heart started to pound and she felt adrenaline rush through her body. The virus was of type H5N1. It was a flu strain that should never have infected a human being. Even worse, Cox was told, the person it infected was a child, a three-year-old boy in Hong Kong. And he had died.

It was August 1997. Jeffery Taubenberger had just published his initial analysis of the 1918 flu virus genes that he had extracted from Private Vaughan's lung tissue. But it was too soon to say what had made the 1918 flu virus so lethal. The flu virologists in Cox's group had no way of knowing whether the virus that had killed the Hong Kong boy shared deadly features with the

1918 flu and whether, like the 1918 virus, the Hong Kong virus would sweep the world, spreading a swath of devastation in its wake. The question leaped to Cox's mind: Was this the first sounding of a fatal pandemic? On the other hand, she realized, it might instead be a repeat of the sort of false alarm raised by the soldier who died of a swine flu virus in 1976.

Cox spent long afternoons in conference calls with her staff and other scientists discussing what to do. She tossed in bed through several long, sleepless nights worrying. The flu virologists of the world could not afford to make a mistake.

Until that moment, there was little reason to panic over the viral infection. Yes, the boy's illness had been frightening, but his doctors were not even convinced that a virus had killed him.

The boy had died on May 9, in a hospital, tethered to a respirator. He had been healthy and entirely normal, striding into his preschool class, playing with his friends, with no illnesses other than the usual runny noses and earaches that plague young children. Then, one day in early May, he got a respiratory infection that quickly turned into viral pneumonia. Soon he was hospitalized, unable to breathe on his own. His doctors diagnosed viral pneumonia complicated by Reye's syndrome, a disorder that sometimes follows viral infections like influenza or chicken pox. It is a rare disease that strikes children and teenagers and it can be fatal. The patient's brain fills with fluid, creating so much pressure inside the skull that the brain starts to compress the delicate nerves at its base, the brain stem, that control breathing and heart rate. When the brain stem is damaged, Reye's victims die.

So even though the child died within days of becoming ill, it was not clear what had killed him—a viral illness or Reye's syndrome. Nonetheless, the hospital staff was frightened and saddened and sought answers. What kind of virus had precipitated this death? How could a vibrant and robust child have become so sick and died so quickly? The child's doctors sent washings from his throat to a government virology lab for analysis. The samples, tests revealed, contained just one kind of virus, an influenza virus. Then the difficulty occurred. Try as they might, the lab scientists could not identify the strain of influenza.

Labs, like the one in Hong Kong that analyzed the throat washings from the three-year-old boy, keep a set of antibodies that recognize the most common types of the viral surface proteins—hemagglutinin and neuraminidase—that define flu strains. Scientists mark the antibodies with chemicals that will glow if the antibodies latch on to a flu virus. Then they swish a solution of these antibodies onto a petri dish where the flu virus is growing. If the antibodies hook up with a flu virus, the petri dish's contents will glisten red.

The boy's virus never elicited such an effect. Even though the lab workers tested the boy's flu virus with every antibody they had, they came up empty-handed. Nothing matched.

The lab workers in Hong Kong were not alarmed. After all, their set of antibodies would attach to the most likely strains of flu viruses, but they by no means had a complete set of antibodies. They passed the sample on to a specialty lab in Rotterdam for further testing.

Since the Hong Kong scientists did not convey a sense of urgency, the Rotterdam scientists simply put the specimen on their list of things to do. In July, they sent some of the sample to Cox's group at the Centers for Disease Control and Prevention.

"They didn't send us any paperwork indicating that it was anything unusual," Cox said. "As far as they knew, it was just another influenza virus." And so her group put the specimen in its queue. "It was processed along with other viruses," Cox notes.

That meant that it was a month before Cox's group got around to looking at it. The Atlanta lab, one of four in the world that keep a lookout for emerging flu strains, is awash in flu samples, receiving several thousand specimens each year. It's part of the global flu surveillance network that allows virologists to discern the first signs of the next year's predominant flu strain—in time to start making vaccines—and that lets them keep an ever-vigilant watch for new strains of flu.

The surveillance network evolved over the years so that now, in the United States, about 110 local influenza centers collect flu viruses in their own regions and determine their type. About eighty-four countries are linked in an international network. The group from the Centers for Disease Control and Prevention looks at flu viruses submitted from all of them. Some places send in just a subset of their viruses, a representative sample. Others send in everything they come across.

"We ask for viruses isolated early and late in the flu season—that sometimes gives us a clue to what is going to happen in the season that is coming or the next season. We ask for isolates from the peak of flu activity and we ask for typical strains and unusual strains," Cox said.

The Hong Kong virus came through that surveillance network.

As a reference lab for influenza, Cox's lab—and the one in Rotterdam—had sets of antibodies to flu strains that no one ever expected to see in humans. These were strains of flu viruses that infect birds. While very occasionally a bird flu mutates and kills birds, for the most part bird flus are entirely benign. Instead of infecting cells of the lung and causing sickness, the virus lives peacefully in cells of the birds' intestines, causing no symptoms. In theory, a bird flu could not infect a human because the virus should require cellular enzymes found in bird intestinal cells but not in human lung cells. Yet if, against all odds, a bird flu virus was infecting people, it would have hemagglutinin and neuraminidase proteins that had never been seen before by a human being. No human would be immune to such a virus. The whole world was at risk.

Worse yet, if a bird flu did jump to humans, and if the event happened in Asia, the scenario fit all too well into a chilling story developed by two leading flu virologists, Dr. Robert Webster of St. Jude Children's Research Hospital in Memphis and Dr. Kennedy Shortridge of the University of Hong Kong.

Webster proposed that the worst flu pandemics, the one in 1918 being at the far end of bad, start with a bird flu. But before it can infect a person, it has to be humanized—that is, to change in a way that would allow it to keep the birdlike features that make it so infectious and yet acquire human flu-like properties that would allow it to grow in the lung cells of a human being. That crucial step, Webster said, typically takes place in pigs. Pigs bridge the gap between birds and humans—both bird flu strains and human flu strains can grow in pigs' bodies.

An unfortunate pig that happens to be infected with both a bird and human virus at the same time can become a mixing bowl, with the genes from the two types of flu viruses recombining in its cells to form a new hybrid virus that can infect humans but has some genes from the bird flu, genes that make the newly emerging virus more dangerous than any that had been around before. Thus the stage can be set for a worldwide pandemic.

As evidence for his hypothesis, Webster posited that the 1918 virus probably started out in a bird, moved to a pig, and then infected people, which is why those who lived through the epidemic had antibodies to swine flu. Moreover, the only two pandemics in which flu viruses had been isolated, the "Asian" flu of 1957 and the "Hong Kong" flu of 1968, involved virus strains that seem to have come, indirectly, from birds. (Earlier pandemics took place before virologists knew how to test flu strains and there were no subsequent pandemics.)

Kennedy Shortridge elaborated from there. Asia, he said, is the influenza epicenter. The virus thrives in ducks, in particular, that are omnipresent in southern China. Those birds have served as a reservoir for dangerous viral strains that have become converted into human flus because of an ingenious system devised by Chinese rice farmers that inadvertently ensures that the flu strains have plenty of opportunity to jump from ducks to pigs to people.

As early as the seventeenth century, these farmers discovered a way to keep their rice crops free of weeds and insects and, at the same time, keep a flock of plump ducks around for food. While the rice is growing, they put ducks on the flooded fields. The ducks eat insects and even weeds, but do not touch the rice. When the rice starts to blossom, the farmers remove the ducks from the rice fields and put them on waterways and ponds. After the rice is harvested, the farmers put the ducks back on the dry rice fields, where they eat the grains of rice that have fallen to the ground. Now the ducks are ready for slaughter.

The problem, however, is that the farmers also keep pigs that live alongside the ducks. And so, Shortridge said, "when you domesticate the duck, you unwittingly bring the flu virus to humans."

Shortridge notes that influenza epidemics always seem to start in Asia—in southern China in particular, exactly the place where the rice-duck-pig system is in place. "Historical records always refer to this part of the world," he said.

Now, as Cox looked at the lab records from the little boy who died in Hong Kong, she knew she was seeing an unprecedented and possibly horrifying event. Here was a flu virus. It came from Hong Kong. It was a bird virus, but unlike any other bird flu virus ever known, it seemed to have skipped the pig step altogether since it had hemagglutinin and neuraminidase proteins that are characteristic of bird, but not pig, flus. It infected a three-year-old boy. And it killed him.

• • •

"Protect the lab workers," Cox thought. If there was any chance that this virus was deadly, like the flu of 1918, it could no longer be treated like any other flu virus. Ordinarily, influenza is deemed a "biosafety 2" agent, meaning that laboratory workers do their experiments under hoods that suck air upward rather than into the faces of the scientists. Since flu is a respiratory virus, this sort of protection stymies most viral infections.

"Of course, we have limited access to our buildings," Cox said. "And laboratory personnel who work on flu viruses use gloves and wear lab coats." But influenza viruses are everywhere, constantly being passed around from person to person, and so the Centers for Disease Control and Prevention does not deem working with them an extreme danger. "You are much more likely to get it in the community than in the lab, that's for sure," Cox said.

But if the Hong Kong virus could be another 1918 flu, the scientists had to go to the next level of protection. "We moved immediately to a biosafety level three-plus containment facility," Cox said. That meant extra layers of protective clothing and hoods with filtered masks that kept viruses out. The lab workers were garbed as though they were handling a deadly plague, which is what Cox feared this virus just might be.

On the other hand, it could all be a mistake. Perhaps there was a mix-up somewhere along the line, perhaps the boy had not died from a bird flu. When something that unusual occurs, the most likely explanation is that there has been a mistake.

"We wanted to be sure that the virus was really from a person and that it didn't come from a contamination," Cox said. "There were a lot of skeptics."

The first order of business was for Cox's lab to redo all of its tests. Fortunately, the lab in Hong Kong still had portions of the fluid washed from the boy's throat, allowing the investigators to test the sample anew. The result was the same: H5N1. A bird flu.

As an additional check, the virology lab in Rotterdam happened to have tested the boy's virus at about the same time as the

Atlanta lab did its test and it had come to the same conclusion. This was an H5N1 virus.

Still, it was possible that a bird virus had contaminated the original sample, a circumstance that could fool Cox and the scientists in Rotterdam into thinking that the boy had been infected with a bird flu. Every scientist who works with viruses learns to suspect contamination. It can occur so easily, with a virus that just happened to drift into a sample of cells, landing on them, taking hold, and replacing the virus that scientists had thought they were studying. In fact, it seemed more likely that the specimen had been contaminated than that the boy had been infected with an H5N1 flu strain. After all, said Cox's colleague, Dr. Keiji Fukuda, an infection of a human being with a bird flu "had never been reported before."

The other possibility, of course, was that the finding was real. And then, Fukuda said, the question became "Was this unique, was this one person? Or was there a new epidemic brewing?" Could it even be the start of something as awful as the 1918 flu? It was hard to put that possibility out of mind.

The only way to decide if the specimen had been contaminated or if the boy really had been infected with a bird flu was to send a team of scientists to Hong Kong to investigate. The group included scientists like Fukuda, from the Centers for Disease Control and Prevention, leading academic scientists like Webster, and investigators from the World Health Organization. They arrived in Hong Kong in August, shortly after Cox had seen the chilling results from her lab's test of the boy's throat washing.

They came with a list of questions, detailed and precise. Their mission was to leave no possibility unexamined, to make no mistakes. Even if their examinations appeared to be overly fastidious, they knew they had no choice. If there was an alternative reason for the bird flu virus being found in the throat washings of the dead child, they needed to uncover it.

The questions moved the scientists backward in time. First, they asked, Was there evidence that the bird virus was a laboratory contaminant? If not, then was there evidence that the virus actually caused the boy's illness? In other words, might something else have made him ill and the bird flu have been only coincidentally present? If he really was infected with a bird flu, where did the virus come from? How did the child become infected? And, the most dreadful of the list of questions: Was there evidence of widespread infection of others with that same bird flu virus?

"We felt that this was the best way to get at the real question: Is there evidence that this is a harbinger of an epidemic?" Fukuda said.

They took the questions one by one and tried to answer them.

To see if the bird virus was a contaminant, the group went to all of the places where contamination could have occurred, starting at the hospital where the boy had breathed with the aid of a respirator. They interviewed the hospital staff, asking whether any of them were ill at the time. They investigated the intubation, questioning the fate of the tube that was put down the boy's throat for the respirator. Were other patients intubated at the same time and had any of them had a similar illness? They inquired if any of the staff lived near poultry farms. When the child was intubated, was anything dropped on the floor? The group learned how the throat washing was carried to the laboratory refrigerator where it was stored and how it was transported to the government virology lab that did the initial testing. They asked if there were new people who came into the lab and if there were reagents in the lab that might have been contaminated. They investigated whether the government lab that tested the boy's throat washings had previously been working with animal viruses.

The scientists found nothing amiss.

"We didn't find any breaks in technique. There were no other people in the intensive-care unit with respiratory illnesses, there was no unusual absenteeism or illnesses among the staff. The government labs were very clean, very well run, very well organized," Fukuda said. From start to finish, the group concluded, the Hong Kong investigators had done everything right and had every procedure in place to minimize the chance that the sample could become contaminated.

The international team also found other evidence that the sample almost certainly was not sullied by a stray bird flu virus. On the same day that the Hong Kong scientists began growing the virus from the boy's throat in their laboratory, they also were working with eighty-five additional specimens from other patients. Four contained influenza viruses, but all four were of a common human strain. If a virus is contaminating a lab, it tends to alight on more than one specimen. So the fact that only the boy's specimen contained a flu virus that was H5N1 indicated that the specimen probably was not contaminated in the Hong Kong lab.

But to be even more certain, the investigators looked at one more bit of evidence to determine where in the sample the virus was growing. They added fluorescently tagged antibodies to the protein H5 to the throat washing and found that the antibodies attached themselves only to respiratory endothelial cells, those that come from the lining of the lungs. If the bird virus had been a contaminant, the scientists expected to find it floating free in the sample, not sequestered in lung cells.

"When we put all this together, it really told us, in about a week, that contamination is a very unlikely explanation," Fukuda said.

• • •

The group moved on to the next question: Was the flu virus related to the boy's illness or was it an innocent bystander? The only way the scientists could address this question was to review the child's medical records and to talk to the doctor who took care of him. When they did so, they soon learned that the boy did not have any other diseases. "He was a normal, healthy boy," Fukuda said. They also learned that his illness exhibited all the signs and symptoms of influenza. It had been an illness that "came out of nowhere," Fukuda said.

So if the bird virus did not arise in the boy's throat washings because of contamination and if all indications were that the boy had been infected with an influenza virus and that that virus made him ill, the investigators were left with the next question: Where did the virus come from? How did the three-year-old boy become infected? Was it through direct contact with a bird or was it through contact with another person who had that flu?

One way to find answers was to look in detail at the H5N1 virus, a task taken up by researchers at the Centers for Disease Control and Prevention in Atlanta. There were two possibilities: Either the virus was purely a bird flu. Or it was a mixture of bird and human flu, which might mean that it had grown in humans and adapted itself.

The antibody studies that had identified the virus as H5N1 could not answer this question since they gave only the grossest identification, looking solely at the hemagglutinin and neuraminidase proteins. To learn the origins of the virus, molecular biologists had to examine different genes in greater detail. The work showed, in the end, that the H5N1 virus was purely bird. The sequences of its genes were like ones seen before in birds, but never before in viruses that infected humans.

Meanwhile, in Hong Kong, the international group was continuing its investigation and had found a possible source for the virus: a terrible flu outbreak in chickens. Several months before the boy was infected, the Department of Agriculture and Fisheries in Hong Kong had identified a dreadful influenza-like illness in chickens, called fowl plague. The infection had struck three farms in the Hong Kong area, killing every chicken on the first farm and about three-quarters of the birds on the other two farms. Five thousand birds were dead. Even more chilling was the scientists' discovery of the viral type. It was an influenza, type H5N1. Perhaps the boy had become infected because he had petted a baby chick or perhaps he had gone to a chicken farm and touched or inhaled bird feces that were teeming with the flu virus.

But the mystery continued. Had the boy gone to any of the affected chicken farms? the investigators asked his grieving family. No, they said, he had not. Were there poultry stalls in the boy's neighborhood? There were not. Were there bird feces on the ground near where the boy lived? The scientists combed the area but found that it was clean. Had anyone else in the boy's family been sick with a flu? Had, perhaps, his mother or father visited a poultry farm, become ill, and passed the illness on to the child? The answer was "no."

There was one other possibility: The boy had attended a preschool, and shortly before he became sick, the school had obtained a few baby chickens and a few baby ducks. The chicks and ducklings were not available for study because one of the baby chicks and two of the baby ducklings had died within a couple of days of being brought to the school and the other chick had disappeared. Were there traces of an H5N1 virus left behind on the schoolroom floor? Could the virus be traced to those pet birds? Carefully, the team of investigators scraped dust off the classroom floor and off the grounds of the nature court where the chicks and ducklings lived and sent the dust off to a lab to be tested for H5N1 viruses. Three months later, they got the results. The virus was not in the school.

Were other children in the school ill? Could another child in the school have gone to one of the poultry farms, been infected with the virus, and passed it on to the child? The group investigated. "We found no unusual illnesses in the other children or the staff" at the time when the little boy got sick, Fukuda said.

The investigators had to conclude that they simply did not know how the child got infected with an H5N1 flu virus. That led them to the next question, and the most important of all: Was there evidence that other people were infected with that bird virus? Was an epidemic brewing?

"We were really concerned that we were in the middle of an outbreak," Fukuda said. The scientists went to the Hong Kong Department of Health, which keeps track of respiratory illnesses among people who visited nine outpatient clinics, to examine records of illnesses. The message was nothing if not reassuring. The Health Department had seen no unusual patterns of influenza or respiratory illnesses. They had tested 4,000 specimens for respiratory viruses and only the little boy's throat washing had contained an H5N1 virus.

Could the infection have come from a neighboring country? Fukuda traveled to China, spending a week there and speaking to public health officials. Once again, he learned that there had been no unusual outbreaks of respiratory illnesses.

Nonetheless, the group was not ready to say that the virus was not a problem. Suppose the child had spread the bird flu virus to other people? Then, the investigators asked, who might those other people be? The answers were obvious: health-care workers, family members, and schoolmates. There also was a possibility that laboratory workers who had handled the sample could have caught the virus. Since the virus had been on chicken farms, it was possible that farm workers had been infected. The telltale footprints of a past viral infection are in the blood; antibodies are formed as the immune system staves off the viral attack. That meant that the group had to start checking people's antibodies.

"We collected several hundred blood samples," Fukuda said. They found four infections with an H5N1 virus, all from groups of people who were likely to have been infected: a lab worker, a worker on a poultry farm, a child in the boy's preschool class, and a parent of a different schoolmate. No one in the boy's immediate family had been infected. And no one who had had no reason to be infected—healthy blood donors and children whose blood had been stored because they had participated in an unrelated vaccine study—was infected.

The group, finally, was satisfied. In September, when the investigators concluded their study, they reported that it appeared the boy really was infected with an H5N1 flu virus. But the virus did not seem to be spreading in the human population. Even though a few people may have been infected with an H5N1 virus, there did not seem to be an incipient pandemic.

They recommended that Hong Kong increase its surveillance and then they went home, relieved and satisfied that they had done a thorough job and that the situation was under control.

Nancy Cox, back at the Centers for Disease Control and Prevention, still had a nagging worry. That child's death in Hong Kong was, after all, a frightening episode. She told herself not to be alarmist. "We kept looking back and thinking, 'Okay, this case occurred in May. It's now September and we haven't had an additional case. That's a good sign.' We were thinking it's probably an isolated incident."

Then, right before Thanksgiving, Cox got a phone call from Hong Kong. "We were informed that additional cases had occurred," she said. She was shocked. "Things snowballed from there," she adds. Now that the virus had appeared again, the bone-chilling fear came back. Was this 1918 all over again?

"What we were really concerned about was that we were seeing the tip of an iceberg," Cox said. "There might be many, many more cases that are undiagnosed. With influenza you have a pyramid. At the top you have the number of deaths and that's small compared to the number of hospitalizations. The number of hospitalizations is small in relation to the number of asymptomatic infections." How many people might already be harboring an H5N1 flu virus? "The bottom line was: Was this virus able to spread from person to person?" Cox asks.

Dr. John LaMontagne, the deputy director for the National Institute of Allergy and Infectious Diseases, was in India at the time, on an official visit with the Secretary of Health and Human Services. He checked his E-mail and saw there, to his horror, a note from a program officer at the Centers for Disease Control and Prevention informing him that the H5N1 virus had reemerged.

"I was very worried," LaMontagne said. "I had remembered that there was a case in May, but this was a six-month interval. The fact that it had come back after six months was very worrisome to me. It suggested that the virus was circulating somewhere, in humans or in animals. There was a possibility that we would have a much more serious problem."

There was no time to waste, LaMontagne thought. His group had to get started making a vaccine. The first batches would go to the lab workers, to protect them. But the scientist also had to lay the groundwork for making huge amounts of a vaccine, enough to protect most of the world, if necessary. They had to work with drug companies and they had to prepare for the worst.

In the meantime, epidemiologists resumed their investigation in Hong Kong, but now it would be even more elaborate. The international group that went to Hong Kong that autumn was much bigger, with seven people from the Centers for Disease Control and Prevention and a virtual army of experts from Hong Kong.

The number of cases kept mounting. From November until the end of December, eighteen people were hospitalized, eight ended up on respirators, and six died.

"The cases we were seeing were such severe cases," Fukuda said. "We were seeing young kids get hospitalized. We were seeing young adults get hospitalized, end up on respirators and die, and people who were generally healthy."

Most of the flu victims were children, Fukuda adds. But "the striking thing was that of the people who became seriously ill, ending up on ventilators or dying, most were people older than eighteen. That was chilling." It was, in fact, an unusual pattern of death, and it had not been seen since 1918.

"We were seeing young people becoming infected and primarily young adults who were at the highest risk for dying. That was very sobering," Fukuda said.

At that point, in the late fall, public health officials were conducting two parallel investigations. One group studied whether the bird flu was spreading among the poultry in Hong Kong and, if so, how quickly and how far. The other asked what was happening in humans. They wanted to know the risk factors that made it more likely for people to be infected with the H5N1 virus. Did the virus still appear to be spread through contact with birds or was there some other way of spreading it? They wanted to know if there was now evidence that the virus was spreading from person to person. Was there evidence that the virus itself was changing, adapting itself to spread more easily from person to person?

The ultimate question was the most frightening. "How likely was it that we were on the precipice of an epidemic? How likely was it that something was starting up?" Fukuda wondered.

It still was not known if the new flu virus shared major features with the 1918 flu. Taubenberger was working on his snippets of the 1918 flu virus, but there was little hope that he would know in time what had made that virus so deadly. The H5N1 virus had come just a little too soon.

At the Centers for Disease Control and Prevention and at the National Institutes of Health, the work on a vaccine proceeded with haste. There was no time to wait for the investigation in Hong Kong to be completed. If the H5N1 virus was spreading from person to person and if it was a killer flu virus, they needed to be prepared with a vaccine. By the end of January, they had enough to immunize lab workers.

Meanwhile, the scientists began searching for a virus that had the H5 and the N1 proteins on its surface—so a vaccine would protect against the H5N1 flu—but, because of changes in other genes, could grow in chicken eggs. LaMontagne beseeched drug companies to start producing an H5N1 vaccine, but the companies were reluctant, he said, because they feared that the virus might start to spread in their flu vaccine manufacturing plants and contaminate their flu virus cultures. "They were afraid it would compromise their ability to make conventional vaccines," LaMontagne said. "If we had an epidemic it would be a different matter," he adds, but in the absence of an epidemic, the vaccine makers hesitated. Yet the scientists knew that even if vaccine manufacturing began that moment, it would be nine to twelve months before they could have enough vaccine to protect the nation. And by then it might be too late. "It's a huge logistical problem," LaMontagne said.

While the laboratory scientists worked with the H5N1 virus, Shortridge coordinated an investigation of birds in Hong Kong. He was worried about what might be happening if the bird virus and a human flu virus mixed in a patient, emerging as a new virus of unprecedented potency.

"We were dealing with a very potent virus and this is one of the concerns that really worried us, that if the virus were to reassort in a human with a prevailing ordinary common flu of the time, we could have had a disaster on our hands," Shortridge said.

He immediately thought of the "wet markets" where poultry is sold in Hong Kong. Every day in Hong Kong, crates of chickens arrive from the countryside, millions of chickens. The live fowl are slaughtered before their customers' eyes in stalls that are strategically located, often in the center of the hotel and business sections, where infected people could spread the virus to travelers, sparking a worldwide epidemic.

"The people here like their chickens fresh," Shortridge said. "Hygiene consists of a douse with cold water," to wash away chicken waste. Since the flu virus grows in the guts of chickens, the slaughtering of the birds in wet markets seemed to offer a perfect opportunity for an H5N1 virus to spread to humans.

All through the fall of that year, the chickens had seemed healthy. Then, shortly after Christmas, some of the chickens in the Hong Kong markets started behaving oddly, and began dying. The sickness commenced with a sort of yellowish diarrhea, then the wattles on the top of the birds' heads drooped. In a couple of days, they would fall over dead, hemorrhaging, with widespread organ damage. The cause was an influenza virus, type H5N1. But how widespread was it? How many chickens being offered for sale were infected with that virus?

To find out, Shortridge and other investigators started gathering fecal material from the city's wet markets, taking the samples to the lab and injecting them into chicken eggs to see if a virus would grow. Influenza viruses flourish in chicken eggs—this tried-and-true technique was the same one that Johan Hultin used in the 1950s when he was vainly attempting to grow the 1918 flu virus from tissues he had removed from frozen corpses in Alaska. But while Hultin failed to get a flu virus to grow from the tissue he recovered, Shortridge, with his bird specimens, succeeded spectacularly: one out of every five chickens appeared to be infected with an H5N1 flu virus. "So this indicated, more or less straightaway, that the chicken was the principal source of virus for humans in Hong Kong," he said.

He also realized where the infected chickens were coming from: China. That country supplies 80 percent of Hong Kong's chickens, shipping 80,000 to 100,000 birds a day into the city. The chickens may be housed in markets for days before they are transported to the wet market stalls to be slaughtered before customers' eyes. The practices leave a perfect opportunity for a flu virus to spread among chickens and then to infect humans.

Moreover, the birds were coming from Guangdong Province, the exact place in southern China where the 1968 Hong Kong flu pandemic began, Shortridge said. He was haunted by the thought that another 1918 pandemic might be starting and that the decisions he and the other scientists made could determine whether the world was ravished by disease and death or whether a disaster was stopped in its tracks.

"It was absolutely terrifying. You could feel the weight of the world pressing down on you," Shortridge said.

While Shortridge and his team investigated the chickens, Fukuda and his group were asking about the spread of the virus among people. They repeated many of the steps they had taken when they investigated the three-year-old boy's fatal viral infection just a few months earlier.

One key investigation involved hospital workers. If those who took care of the flu patients had antibodies to the H5N1 virus, that would indicate that they had been infected. For comparison, Fukuda's group took blood from hospital workers in other areas of the buildings who had not been anywhere near the flu patients and from hospital workers who were from hospitals that had not admitted any patients with an H5N1 influenza. "We kept in mind that if the spread was efficient we would be seeing a lot of illnesses in the staff," Cox said. "Of course, everything is confounded if there is a regular influenza spreading in the community. In Hong Kong, there was another influenza spreading. There were lots of patients coming in and we had to screen them. There was a heightened concern, everyone with a respiratory illness was concerned that they might have the bird flu." To respond to the near-panic in the city, some hospitals instituted twenty-four-hour flu services, screening people around the clock.

At the same time, the investigators interviewed friends and family members of flu victims. They took samples and tested them for H5N1 viruses and for antibodies to the virus. But they also did something more sophisticated.

Fukuda explained: "We instituted a number of analytic studies. In one form or another we were comparing sick people to not so sick people, trying to find out what's different about the sick people. We looked at people who had gotten sick and compared them with people who were not sick. And we looked at groups who were exposed to one of the people who had become ill and compared them with people who had not been exposed."

It was a huge job, Fukuda observed, and the analyses were more difficult than they might seem at first glance. The problem was that people could have been infected because they came into direct contact with someone who was ill with an H5N1 flu infection, or they could have been infected because someone who was ill passed the virus on to someone else who passed it on to them. In one carefully documented case, at least 7 percent of people infected got a symptomless flu—they were infected and infectious but they showed no outward signs of being ill and did not realize that they were harboring a flu virus. A final possibility was that people could have been infected because they were exposed to poultry. "The studies were set up to help us tease through those possibilities," Fukuda said.

The scientists were working as quickly as they could, aware that panic was growing. "Everybody was frightened," Shortridge recalled.

"There was an unbelievably high level of anxiety in Hong Kong," Fukuda said. "There was tremendous, really tremendous scrutiny worldwide."

It took time to complete the studies; some, in fact, took a year. But as the evidence started coming in, the picture was consistent: There was no evidence that travel or eating poultry or visiting aviaries or having other kinds of pets increased a person's risk of getting the H5N1 flu. "The single strong risk factor that fell out was being exposed to live poultry in the week before becoming ill," Fukuda reasoned.

With the two lines of investigation pointing to the wet markets, there was just one conclusion. The chickens must be slaughtered—all of them.

One day, as Shortridge strolled through a wet market, he came upon a terrifying sight. "We saw a bird standing up there, pecking away at its food, and then very gently lean over, slowly fall over, to lie on its side, looking dead. Blood was trickling from the cloaca. It was a very unreal, bizarre situation. I had never never seen anything like it." And then he witnessed it in another chicken and another chicken. "We were looking at a chicken Ebola," said Shortridge, referring to the horrific viral hemorrhagic fever that is one of the most deadly viral infections of humans.

"When I saw those birds dying like that, it really hit home what might have happened in the 1918 pandemic," Shortridge said. "I thought, 'My God. What if this virus were to get out of this market and spread elsewhere?' It was an unbelievable situation, totally frightening. My mind just raced."

The killings began on December 29, 1997, with an announcement by Steven Ip, the Secretary for Economic Services. "We will start destroying all the chickens in Hong Kong Island, Kowloon Island, and the New Territories," he announced.

That meant killing more than 1.2 million chickens from 160 farms and more than 1,000 retailers and stalls. Some chickens would be killed by their owners, others would be carted away by the government to be gassed to death. Then their carcasses would be disinfected and buried in landfills.

It was an unforgettable sight. In one market, called Fai Chai Lam Cheung Kai, workers began at eight o'clock in the morning, pulling chickens, ducks, pigeons, and quail out of their cages stacked in tall piles. Using sharp knives, they slit each bird's throat and discarded the carcasses in plastic garbage bins. It took ten minutes. The result: one hundred birds were dead.

In just a day, 770,000 of the territory's birds were killed. The rest were dead by the evening of the second day. Even so, Hong Kong officials said that the process took longer than it could have. Moreover, there was a shortage of gas canisters for mass destruction of chickens. And although the Department of Agriculture and Fisheries requisitioned 1,300 employees to help, most had jobs like dog catcher and park warden. "Most of our staff members had never encountered a live chicken," said Lessie Wei, the director of agriculture and fisheries for the territory. "There was a learning process. Some are now experts in killing chickens."

The chicken markets in Hong Kong were shut down for a month and the Hong Kong government began quarantining birds from China to be sure they did not have the H5N1 flu when they entered Hong Kong. The government insisted that chickens no longer be housed in wooden cages. Now they are kept in plastic cages that can be disinfected more easily. The situation seems to be under control. Over 200,000 chickens have been tested and none have had the H5N1 virus.

Shortridge and Webster, in the meantime, are developing a surveillance system for flu viruses in Hong Kong's birds and pigs. So far, Webster said, "there is no evidence of H5N1 in this part of the world."

Back in England, observed John Oxford, the British virologist, hospitals had put "pandemic plans" in place. Anyone who arrived in Heathrow Airport from Hong Kong with a respiratory infection was quarantined, just in case. The H5N1 scare, he said, "was a dress rehearsal."

Nancy Cox is relieved, but still wary. The story is a triumph of epidemiology because what might have been a deadly pandemic was stopped in its tracks.

"In Hong Kong there is typically a peak on influenza activity in February," Cox said. "Our greatest fear was that some individual would become dually infected with an H5N1 virus and a human strain. Then you would get a hybrid. Our fear was that through a process of genetic shuffling a virus could emerge that could replicate in humans efficiently and be transmitted from human to human and it could be as dangerous as the bird flu."

That bird flu episode, Cox concluded, "was a wake-up call for a lot of people who had forgotten how devastating influenza can be. I think it was important for all of us to take heed and get our pandemic planning efforts reorganized."

# 9

# FROM ALASKA TO NORWAY

In the spring of 1996, while Jeffery Taubenberger and Ann Reid were still utterly frustrated by their failed attempts to get viral genes out of Roscoe Vaughan's lung cells and a year before the little boy in Hong Kong would die of the flu, another kind of drama was unfolding. It involved Kirsty Duncan, a grimly determined young woman, unknown to the ranks of flu specialists, who had had no special training in virology or epidemiology. Instead, she was driven by a passion to solve the mystery of the 1918 flu. In May 1996, Duncan trudged along a gravel path on the side of a mountain on a remote island of Spitsbergen, less than 800 miles from the Arctic Circle, to visit a lonely gravesite. The earth was covered with snow, the air chill with winter, although the long hours of daylight signaled spring. The area was bereft of trees, allowing Duncan to see the cemetery clearly in the distance, rows of white crosses on the mountainside.

"I slowly walked up the valley side, up the steep slope. I knew they were in the last row," Duncan said. Then she was upon them: six white crosses and a headstone. Buried there, in the frozen land of a tiny island, were the bodies of seven young men. They had come from Norway for mining jobs, crossing the cold Norwegian Sea in September 1918. But they had become ill on the journey and died of the flu before they had a chance to step into a coal mine.

Duncan stood beside the graves, reverent and solemn. The miners had been struck down in the prime of their youth; the youngest was eighteen and the oldest was Duncan's age, twenty-eight. Duncan had found their graves after all these years, and as she stood on the icy ground she realized that the miners' bodies might still be frozen with the terrible flu virus intact inside their tissues. If there was any hope that the 1918 virus might be found, it resided in the bodies of these young miners that had lain undisturbed for so long.

She had struggled with the ethics of even asking to disrupt their graves. "I believe a cemetery is a sacred place, you don't disturb someone's final resting place," Duncan said, with a catch in her voice. And yet, if she could find the 1918 flu virus, the work could help scientists to make vaccines and test antiviral drugs; it could help protect the world from another catastrophe if that virus returned. "It's an impossible decision. It's a decision I fought with every day."

Agonizing over the moral thing to do, Duncan turned to her parents and her family for counsel. In the end, her father helped her decide, telling her: "Kirsty, if I held the secrets to this deadly disease, I would hope that someone would come along and unravel them."

And so Duncan had traveled to the tiny town of Longyearbyen in the archipelago of Svalbard in the Norwegian Sea to visit the miners' graves.

Duncan was a geographer, at the University of Windsor and the University of Toronto in Canada, who studied climate change and how it affects human health. She was a striking figure, tiny with waist-long brown hair, a runner, and a devotee of Scottish dancing. Her life changed in 1992 with a casual conversation with her former husband.

"I was married to a pediatrician and we had been talking about the 1918 flu and I said, 'Well, I'm going to read Alfred Crosby's book,'" Duncan recalls, referring to *America's Forgotten Pandemic*. "Initially, I wanted to explore the link between climate and flu."

As she read the book, Duncan became horrified by the flu's devastation. She solemnly recites, by memory, the statistics: Half a million died in the United States, 19,000 in New York City; half a million sickened in Quebec; 14,000 died in Quebec.

"The funeral directors couldn't keep up with the demand for funeral cars and so they converted trolley cars. A trolley car in Montreal could carry ten coffins," Duncan said. The images were searing. "I came home to my family and I said, 'I'm going to find out what caused the disease.'"

She began with an electronic database, Medline, that contains articles from medical journals, pulling every article she could on the flu. Taubenberger's name did not appear; he had not yet published his paper. Then she started calling influenza experts and asking them, Are there any archival samples? Does anyone have tissue from 1918 flu victims that might contain the virus? "I was told, 'No. They're all gone,'" Duncan said.

Unwilling to give up, Duncan asked herself how else she could get an answer. "My background is in geography and anthropology. So I thought, What would preserve biological tissue?" The answers, she said, are extreme dryness, extreme wetness, and extreme cold. "I thought the greatest chance for success would be to look in cold areas." She needed to find flu victims who died and were buried in permafrost.

Duncan began her search with Alaska, a territory where the 1918 flu had struck with a particular vengeance. Crosby had written that while the virus swept the world, it had decimated the villages of Eskimos, killing as many as 90 percent of the young adults in villages that it entered; only quarantines and a deliberate policy of some villages to cut themselves off from all contact with outsiders saved them. She reasoned, like Johan Hultin forty years before her, that if she could find Eskimos who had died of influenza in 1918 and had been buried in regions of permafrost, she might be able to locate the flu virus.

She began by writing to the Alaska Bureau of Vital Statistics for the records of everyone who had died in 1918. The response was overwhelming, but not helpful. Duncan got back 2,000 death certificates but she did not know which flu victims to focus on. "I could not correlate flu victims with permafrost," she said. "The U.S. Geological Survey did not have permafrost maps that were good enough to do that."

If Alaska would not work, what about Iceland? Duncan thought. But she reconsidered. "The geothermal energy would not allow for preservation," she concluded.

She thought of Siberia and Russia, but again, no luck. "I wrote to the Russian and Siberian authorities but got no response," Duncan explained.

Duncan turned to the Norwegian islands in 1994. A close friend of hers had led an expedition across a glacier in Svalbard, an archipelago of islands owned by Norway, about 600 miles north of the mainland, across the Norwegian Sea. When he described his trip to Duncan, he mentioned permafrost. "I thought, 'That's it,'" Duncan said. She went to the University of Toronto library to read up on the area and, in particular, to learn about the port town of Longyearbyen. She found just two books written in English, and learned that Spitsbergen had been a center for the whaling trade, with 200 to 300 ships and 10,000 to 20,000 men in the area in the late seventeenth century. Coal mining began in 1906 when John Monroe Longyear, an American, formed the Arctic Coal Company on the island, and within a decade there were six coal mines in the area. The miners and their families lived in the newly built town of Longyearbyen. The town also drew seasonal workers, men who were farmers or fishermen in Norway, who spent the cold winter months in Longyearbyen mining coal.

Duncan took a guess: influenza had devastated Norway. Norwegians had traveled to Spitsbergen for mining work. The flu might have come to Spitsbergen. And if it had, there might be flu victims buried in the permafrost. What Duncan needed were records to point her to influenza victims.

She wrote to the Norwegian Polar Institute, but the news was discouraging. There were no hospital records because the hospital was destroyed in World War II. There were no church records, because the minister didn't come until 1920. There were no government records because they were in buildings that were bombed and destroyed in World War II. "Things were looking desperate," Duncan said. The institute did offer her one ray of hope: there were diaries kept by the coal company.

Duncan called the coal company. "They said, 'We no longer have the diaries.' I just about died," she recalled. Then the company told her that a local schoolteacher in Longyearbyen had them. When Duncan phoned the teacher, Kjell Mork, he agreed to translate them for her and told Duncan about the seven miners who had perished from the flu.

The men had departed from Norway for winter mining jobs on September 24, 1918, sailing for three days before arriving at Longyearbyen on the last journey of the year before the ice made the frigid waters impassable. But the influenza virus was spreading on the ship. Within days of disembarking at Longyearbyen, seven men were dead from the terrible illness. Their bodies were kept outside, where the temperature had been well below zero. Before they were buried, their names and their dates of birth and death were inscribed on the six crosses and one headstone that marked their graves. The practice at the time was to bury the dead in simple wooden coffins, without embalming fluid.

"After two years of searching, now I knew of these young men," Duncan said. But could she find the graves, and, if so, had they been disturbed? She wrote to the minister, who told her that the graves were marked and had not been touched since 1918.

No one could say for sure how deeply the men had been buried—a crucial question. Although permafrost reaches to a depth of 500 meters in some areas of Svalbard, the top half meter to meter of ground melts and refreezes each year. That is the so-called active layer. Duncan contacted archeologists, historians, the governor of Svalbard, and even the local funeral director to discover the burial practices at the time. Bodies typically were buried two meters under the ground, she learned, although there was no guarantee that the practice in Norway was also the practice in Svalbard in 1918. "I was told by the governor's office that Svalbard was a no-man's-land," Duncan said. "There was no law. We just don't know."

On the other hand, there was reason to suspect that the miners might have been buried below the active layer of the permafrost, Duncan reasoned. They had died in October, when the permafrost was melted most deeply, making it the easiest time of year to dig down deep. In other places where the flu had struck, people were so frightened that they dug graves as deeply as they could. And bodies buried in shallow graves in permafrost regions like Svalbard sometimes floated to the surface as the active layer froze and thawed and heaved bodies upward. No bodies had surfaced in the Longyearbyen cemetery, Duncan learned. She thought that the miners would be buried deeply. "They were miners. If anyone knew how to dig in this material, they did," she said.

Duncan had found her flu victims. Now she needed to organize a group to find out if there were intact bodies in those frozen graves and, if so, to exhume them, remove tissue from their lungs and other parts of the bodies, and see if the 1918 influenza virus was still there. She began approaching scientists, telling them of her fantastic discovery and asking them to join her team. And she waited for permission from the Norwegian authorities to go ahead and dig.

In February 1996, she got it: the governor of Svalbard told her that she had contacted all the relevant parties—the Norwegian medical research community, the church council, the bishop, the town council, and the families of the miners—and they had agreed to let her continue.

"I was just overwhelmed," Duncan said. "I'm an outsider and what I was asking permission to do was potentially very hurtful." She was moved by the decision of the families to allow her to open the graves of their relatives. "I wonder how many of us would be willing to grant permission," she said. "I really treat their permission as a very great gift and I treasure that gift."

It was time to start planning the expedition in earnest.

One of the first to sign on was Dr. Peter Lewin, a pediatrician and medical archeologist at the Hospital for Sick Children in Toronto. Lewin describes his work in medical archeology as "an exotic hobby." He has autopsied mummies and has used modern scanning methods, like CT scans, to probe the medical mysteries of pharaohs of ancient Egypt. He had found evidence of the smallpox virus in the mummy of Ramses V. He was part of a team that examined tissues from sailors who died in the 1840s, members of a disastrous expedition to the Canadian Arctic. He had searched for smallpox in bodies buried in the Canadian permafrost, not finding it but convinced that a virus could remain viable in long-frozen corpses. Until he spoke to Duncan, however, he had never thought about the 1918 flu.

Duncan arrived at Lewin's office, telling him about the flu and the miners and the possibility of finding one of the most deadly viruses the world has ever known. Lewin was sold. "We never looked back," he said.

"I was very excited," Lewin said. Duncan's plans "fit in very well with my ideas." He was thrilled to be given an opportunity to help isolate the 1918 flu virus.

The team began to take shape, with experts from a variety of disciplines. It included Dr. Alan Heginbottom, a Canadian geologist; Dr. John Oxford, the British virologist; Dr. Robert Webster, an American virologist; and Sir John Skehel, the director of England's National Institute for Medical Research in Mill Hill.

It was an emotionally complicated affair. Duncan, with her long hair and doe eyes and raw emotions, was undeniably attractive to some of the men on her team. Esther Oxford, John Oxford's daughter, wrote an account of her father's involvement with the group, and said that soon after Duncan and her father met, they started "an affair of fax and telephone which lasted 'too long,' according to Gillian Oxford, my mother. Meals were interrupted, work-days were filled with constant messages, calls and faxes—all of which featured touchy-feely words such as 'overwhelming,' 'truly,' 'deep.'" Duncan told Oxford's daughter, "Your father and I were excited by one another."

And when Duncan's marriage broke up, she called Oxford in tears. By then, Esther Oxford wrote, her father had gotten all of the funding for ground-penetrating radar, half the funding for the actual exhumation of the bodies, had introduced virologist Rod Daniels to Duncan, and had found the Necropolis Company to do the grave digging.

For Duncan, however, there was a huge gap in her preparations for the expedition: after two years, during which time Duncan had spent $60,000 of her own money investigating the possibility of finding the flu virus and planning the expedition, she had never actually been to Longyearbyen. She had never seen the miners' graves. It was with great excitement and some trepidation, then, that she set forth that year in May.

"I was very frightened, so afraid that I would hurt the people living there and the church. How would people feel? That really bothered me," she said. On the second day, she went to see the minister of the church in Longyearbyen.

"I introduced myself and said, 'I hope I have not offended you and the church.'" The minister, she recalls, told her that, no, she had not offended him; hers was important work that needed to be done. Then he asked Duncan if she had been to the cemetery. "I said, 'I have no right to go without your blessing.'" The minister told her she had it and that she should go.

"It was the longest, hardest walk of my life," Duncan said. "It is quite a distance to the cemetery and you can see the white crosses in the white snow."

After her solemn journey she returned to her hotel. Before she left the island, she made a second visit. It was the middle of the night, with the Arctic sun still burning in the thin cold sky. Duncan walked up the path to the cemetery, to the last row of graves, to the graves of the young miners. With her hands she dug seven shallow holes in the snow, one at each grave, and buried seven roses in the seven holes.

Upon her return to Canada, Duncan felt she could be silent no longer about her plans. She needed to tell the public what she intended to do and gauge the world's reaction. "As a member of the public, I would not want to think that some scientist just went up there and did this work," she explained. "I would want to know about it, I would want to know that they did it responsibly and that they did it right. I would want them to debate the merits and drawbacks of the work. If there had been a great hue and cry about the project, I would have stopped it."

But just the opposite occurred. People were touched and excited by her plans and many who knew firsthand how terrible the 1918 flu epidemic had been wrote to Duncan and called her to encourage her in her efforts. "In three years, I have gotten three letters where people expressed concerns and over a hundred letters from people in their nineties, begging me to find an answer," she said. Some of the letters are extremely personal. Some phone calls were so moving that she dissolved in tears. Duncan keeps the letters from flu survivors on her desk "to remind me of why I undertook this, of why I have to keep going."

Meanwhile, Jeffery Taubenberger, working quietly in his cramped quarters at the Armed Forces Institute of Pathology, found out about Duncan through the scientific grapevine. Leading flu virologists, like Dr. Robert Webster of St. Jude Children's Research Hospital in Memphis, were being recruited to help with this fantastic project. Meanwhile, Taubenberger and Ann Reid, people no flu experts had ever heard of, were slaving away trying to pull viral gene fragments out of the slivers of lung tissue they had gotten from the pathology warehouse. And they were having an impossibly difficult time.

It was not easy for Taubenberger to watch the media hail Duncan that spring and to know that he had in his hands human lung tissue containing the 1918 flu virus. But what would be the point of making an announcement? While Duncan talked to scientists and the press about her project, Taubenberger and Reid were still trying valiantly to push the methods of molecular biology to the limit so they could fish the viral gene fragments out of the formaldehyde-soaked scrap of lung tissue from Private Vaughan.

Gradually, that summer, as their luck changed, Taubenberger decided that the best strategy was to stake a claim on the flu virus work, but not by making a public announcement. Instead, he would take the conservative path of academic science: a publication in a leading scientific journal. He needed to let the world know what he was doing, now that he was getting some results. And he needed to raise a question: Should Duncan go forward with a project that might unleash live flu viruses, among the most deadly viruses ever known, when he could get the same information with utter safety?

Taubenberger sent his paper to *Science* magazine that October. "We wouldn't have published our *Science* paper when we did if it weren't for them," Taubenberger said. "We thought, 'We ought to publish what we have now so we can show we are there first.'"

It took until March before Taubenberger's paper appeared in *Science* and it was not until then that Duncan learned of his work. Suddenly, Taubenberger's question was everywhere: Why should Duncan press ahead? Why take a chance that maybe she would release live flu viruses from the miners' bodies when Taubenberger had found this other, safer source of the virus's genetic material?

But Kirsty Duncan was convinced that her project still was valuable. After all, Taubenberger's tissue had been soaking in formaldehyde for nearly eighty years and who knew what might have happened to the viral genes? And it was not clear that Taubenberger even had enough viral material to find the complete genetic sequence of the 1918 virus. Duncan's group wrote about Taubenberger's paper on its Web site: "Results published in March 1997 showed he had managed to recover only partial sequences of some viral genes."

Nancy Cox, chief of the influenza branch at the Centers for Disease Control and Prevention, decided that it was time to bring Duncan and Taubenberger together. She was caught in the middle, having agreed to provide scientific advice to Duncan and discuss the molecular biology of influenza with Taubenberger. Yet Taubenberger and Duncan had never met each other. Maybe if the two sat down and discussed their work, they could agree on the best way to proceed.

The meeting took place in April 1997, just two months after Taubenberger's paper had been published. The two young scientists discovered that they had much in common. Both were musicians, with Duncan playing piano, violin, and bagpipes. Taubenberger promised to send her tapes of the music he had composed. Both were unknowns in the flu world embarking on projects that were breathtaking. Both had devoured Alfred Crosby's book, inspired to press on by his descriptions of what the 1918 flu virus had wrought.

The decision of whether Duncan should proceed turned out to be not a difficult one. One of the criticisms of Taubenberger's work, to which he readily agreed, was that he had only one sample. Perhaps what he thought was the 1918 flu virus was an innocent and innocuous flu virus that was a bystander while the deadly virus killed the soldier. Only with more than one sample from different flu victims could that question be laid to rest. Duncan's expedition could provide more flu samples.

Duncan should go ahead with her plans, all agreed, while Taubenberger, working separately on his own project, continued to determine the sequence of the virus's genes. He agreed to be part of Duncan's team, using the molecular biology techniques his lab had perfected to fish the virus out of the tissue from the miners' lungs. "Everyone felt they should continue on with their project, that finding more than one sample would be useful," Taubenberger said.

A month later, Cox and the others at the Centers for Disease Control and Prevention pulled out of Duncan's project providing only vague explanations: With Taubenberger's project underway, the Spitsbergen project was no longer a scientific imperative. The Centers had limited resources. They had other things to do. "We have so many projects going on that it is impossible to participate," Cox said.

Taubenberger, however, was still officially a part of the team, although he was not in touch with Duncan as he continued with his own work. His research on the flu virus was getting increasingly frustrating. With just a few molecules of viral genes in the sample from Private Roscoe Vaughan, Taubenberger was beginning to wonder if it was even possible to get more than partial genetic sequences.

"There was so little material," Taubenberger said. "It couldn't get any worse." Realizing that Private Vaughan's lung tissue might not be enough, Taubenberger went back to the warehouse samples and screened another thirty-five of them for footprints of the flu virus. One proved positive, the lung tissue taken from the corpse of Private James Downs, a thirty-year-old soldier from Camp Upton, New York, who had died just two hours before Private Vaughan in South Carolina. Private Downs had entered the base hospital on September 23, 1918, and died at 4:30 a.m. on September 26. His lungs were heavy with fluid, exuding a "bloody froth" according to the autopsy notes.

Encouraged by the discovery of the virus-containing tissue from Private Downs, Taubenberger went home for a two-week paternity leave—his wife had just given birth to their second child—confident that when he returned the work with Private Downs's tissue would go forward as planned. In the meantime, his mail accumulated, unread, in his office mailbox.

The envelope was waiting for Taubenberger when he returned at the end of July. It was postmarked San Francisco and it contained a curriculum vitae as well as a few pages that had been photocopied from Alfred Crosby's book on the 1918 flu epidemic. Then there was the letter. It was from a man Taubenberger had never heard of, a seventy-two-year-old retired pathologist named Dr. Johan Hultin.

In careful sentences, Hultin told Taubenberger what he had done in 1951, that he had gone to Alaska and exhumed the still-frozen bodies of flu victims that had been buried in a mass grave in the tiny village of Brevig. He explained that he still had contacts in Alaska and that if Taubenberger was interested, he could return to Brevig and attempt to get lung samples from the corpses of flu victims that were still frozen and buried in that grave. Hultin's caution emanated from his letter. "He was very concerned that we not think he was a kook," Taubenberger said.

Although he was feverish with excitement, Taubenberger responded in the same measured way: he wrote Hultin a letter. Then the two men began exchanging telephone calls, cautiously planning a very different expedition than the one Kirsty Duncan was leading. It would be low-key, paid for entirely by Hultin, with no advance publicity and no publicity at all if he did not succeed in recovering lung tissue from frozen flu victims. Hultin told Taubenberger that he remembered with distaste his return from his expedition to Alaska in 1951. His plane was met by a reporter from *Life* magazine. Now that the media was far more voracious and clamorous, Hultin recoiled in horror at the very thought of sparking a frenzy around a voyage to the isolated village of Brevig. He had to insist: no publicity, no members of the press invited or informed. He would ask the village leaders for permission to reopen the mass grave of the 1918 flu victims and, if they agreed, he would dig up the bodies himself, Hultin told Taubenberger. If he succeeded in getting lung tissue, he would send the samples as a gift to the Armed Forces Institute of Pathology.

The contrast with the Spitsbergen expedition could not have been more stark. Duncan had told the world what she was doing and had spared no effort to be certain that she would ensure the utmost safety, even if she was slowed down, even if, in the end, she did not exhume the bodies of the miners at all. Her project was expensive and was taking place with all deliberateness and with the eyes of the world upon her. Her group had spent endless months and years planning and getting permits and signing up an international team of experts in every conceivable area. Hultin was a one-man, self-funded expedition planned on the fly.

And yet Taubenberger realized immediately that Hultin had more experience in searching for frozen bodies of flu victims than anyone else on earth and he was the only person who had ever succeeded. The sole remaining question was whether it would be worthwhile trying to grow the actual virus if it still could be resurrected from the lungs of one of the corpses in Brevig. That, Taubenberger decided, was a fool's mission.

"The influenza virus is a very fragile virus," he said. "One hour at room temperature and it's dead. So the chances of getting live viruses are virtually zero." But the attempt would require the most elaborate precautions, a biohazard isolation lab where workers wore space suits when they were working with live viruses of the most deadly type. Why even try to schedule such lab work? "I said, 'Take the frozen biopsy samples, put them in fixatives, and we'll look for the viral RNA.'" The viral genes would be intact but the virus itself would not be alive.

Hultin was overjoyed. Ever since his first trip to Alaska in 1951 he had dreamed of returning. As the years passed and Hultin built his successful career as a pathologist, he carefully followed the revolutionary discoveries in molecular biology, waiting for the right moment to mention Brevig and the right person to tell. "Little by little, I began to see the move into molecular genetics." By the 1980s, he said, "I knew that one day I would have another chance." He saw papers on a new method, polymerase chain reaction, or PCR, that would enable molecular biologists to analyze tiny fragments of genetic material and, he realized, perhaps even pull the flu virus out of samples from Brevig, making it irrelevant whether live, or even whole, virus was present. "I read about PCR and I realized now that, hey, this is something, this is coming now," Hultin said. "I had better pay attention."

In March 1997, Hultin read Taubenberger's article in *Science* and he knew his time had come. He tried to make it easy for Taubenberger to say yes to him, realizing that Taubenberger must be skeptical that this seventy-two-year-old retired pathologist could ever succeed in such a spectacular mission. He suggested that he would simply go to Brevig on his own and deliver the lung tissue to Taubenberger, in that way making sure that Taubenberger would not have to even wonder if he would be liable in case Hultin unearthed specimens with live flu virus and unleashed a catastrophic pandemic. Hultin was certain that could not happen. "I didn't even consider it because I knew the virus was dead." But he wanted the risk to be his so that Taubenberger would not encounter legal or bureaucratic entanglements.

"There would be no connection between us other than my delivery of specimens," Hultin said. "I was so anxious to do the work that I wanted to remove all conceivable obstacles. I did not want to lose this opportunity."

He reviewed his plans: He was paying for the trip himself so he did not have to apply for funding. If he asked the government or a company to provide support, they might worry about what would happen if he found live virus and about their potential liability. They might throw roadblocks in his way.

He was telling no one but Taubenberger, so there would be no long and prolonged public debate before he even spoke to the Brevig villagers. He could use his own knowledge of Brevig and his persuasive powers to get permission to dig up that old grave again.

The more he planned the trip, the more "all sorts of little things added up to convince me that I could do it on my own. I didn't ask anybody any questions except Taubenberger, and all I asked him was 'Would you take my specimens?'"

"I was interested in speed," Hultin said. "Influenza might come back while we were waiting for bureaucrats to make a decision."

Taubenberger told Hultin that, by the way, there was a similar expedition being planned by Duncan involving the graves of Norwegian miners. That exhumation had been in the works for four years. Depending on tests of the ground, it was scheduled to take place in the fall of 1998. Accustomed to investigators who do nothing in haste, Taubenberger then asked Hultin when he thought he could go to Alaska. He expected that Hultin would need months, maybe longer to plan the trip. But Hultin's reply took him aback.

"I said, 'I can't go this week, but maybe I can go next week. I'll call my travel agent to fix me up with tickets,'" Hultin recalled.

Taubenberger was so astonished by Hultin's response that he never thought to ask him why he could not go that very week, nor did Hultin volunteer the information. The reason was that he was building a 1,300-square-foot replica of a fourteenth-century Norwegian cabin on some land he owned in the Sierra Nevadas between Yosemite and Tahoe. After twenty-nine years of work, the cabin, with its detailed woodwork and spectacular redwood artisanship, was almost finished. "I had some details left on the cabin and that's why I couldn't go that week," Hultin said. "But I didn't want to tell Jeffery. He might think he was dealing with a nut."

The next week, Hultin was off, without even telling the Brevig villagers why he was stopping by for a visit. He traveled alone, and planned to discuss his mission with the villagers once he got there.

"One cannot discuss a very sensitive matter with Eskimos by telephone," Hultin said. "When it comes to digging in cemeteries and dealing with dead ancestors, one can't do that by phone. You have to go up in a very quiet, low-profile way, and talk to them about it."

Hultin's only phone call before he flew to Brevig was to the city clerk, a man who takes care of correspondence and answers the telephone for the 240 Brevig residents. He asked just two questions: Has anyone been digging in the mass grave? And where could he sleep?

The clerk told Hultin that the grave had been left undisturbed since he last was there forty-six years ago. He also told him that the school had four air mattresses and a kitchen where he could prepare food. And it was heated, a necessity even though Hultin would be arriving in August. Two of the mattresses were taken, being used by men who had come to install a satellite dish in the village. But there still were two left, one for Hultin, and a spare one that could be given to another visitor.

When he arrived, Hultin found himself saddened by life in Brevig, so different from what it had been in 1951. Then the villagers were self-sufficient; many still practiced the ancient skills for whaling and hunting that had been handed down from father to son for generations. By 1997, he said, all that was gone, replaced by a dependence on welfare. Now the village, still isolated beside the chilly gray sea, was desolate, a place of hopelessness rather than pride.

"It is very tragic," Hultin noted. "They exist, they have many children, and the government pays." Families are huge, Hultin claimed, in part because every villager, infants and children included, received money each year—$1,800 in 1996—from an oil company in compensation for the use of their lands. But there was no work, nothing to do.

With their welfare money and the oil company money, the villagers were able to buy snowmobiles and four-wheeled motorcycles. "Every family had one or more of these. I have footage of a snowmobile next to a dogsled," Hultin said. But most of the dogsleds were decaying, relics of an older, simpler way of life.

Hultin, however, did not bemoan the past in front of the villagers. Instead, he set about to get permission to exhume bodies from the mass grave where seventy-two villagers who had died of the flu epidemic in November 1918 were buried. He turned first to Brian Crockett, a missionary who was pastor of the Lutheran church in the village. He and his wife, Ginger, a schoolteacher in Brevig, introduced Hultin to Rita Olanna, an important villager who was matriarch of Brevig's largest family and a member of the village council.

"I showed her my pictures, prints from my previous visit, copies of letters written by Otis Lee, the missionary at the time, copies of mission letters from 1918," Hultin said. "She discovered that her aunt and uncle were mentioned as references for the missionary. She was so delighted to see the name. She had relatives who died in 1918 and were buried in that grave."

Slowly, carefully, Hultin began to describe the reason for his visit. "I said that a terrible thing happened in November 1918 and I am here to ask permission to go back to the grave for a second time. Science has now moved to a point where it is possible to analyze a dead influenza virus and through that make a vaccine so that when it comes again, all of you can be immunized against it. There shouldn't be any more mass deaths."

Olanna told Hultin she understood him and supported him. Pastor Crockett also supported him, and so Olanna called the village council together to decide whether to allow Hultin to proceed.

"We met in the early afternoon. I presented them with the same story I had told Mrs. Olanna earlier," Hultin recalled. The group gave him permission. And, to his surprise, one of the village elders asked him if he needed help. When Hultin had been to Brevig forty-six years earlier, no one had offered assistance. Gratefully, Hultin said that, yes, he could use some help, and was provided with the services of four young men.

Within the hour, on August 19, 1997, Hultin was at the gravesite, clearing the low brush from the area and starting to cut the sod. Then the young men sliced off the earth in foot-long squares, to be replaced later when the work was completed.

With the earth exposed, Hultin and his assistants began to dig, pushing fairly easily through the first two feet of dirt that covered the permafrost below. They opened a rectangle six feet wide and twenty-eight feet long. This time, it was so much easier. "Now we had muscle and picks and spades. Particularly muscle,"

Hultin said. "These were people who were accustomed to digging in permafrost." They had perfected the technique that Hultin describes as "split and shovel, split and shovel."

On the afternoon of the third day, the group came upon the first body. It was a skeleton—no soft tissue remained. Hultin told himself not to be too concerned. After all, this was the area he had disturbed when he had dug into the grave in 1951. It was only natural that those bodies would have decomposed in the intervening forty-six years.

The men were working with no precautions in case they came upon live virus, but Hultin was not worried. "I said, 'The virus was dead forty-six years ago. It's even deader now,'" he explains. He wore gloves, but that was only to keep his hands clean to pick up his camera. Unconcerned, he and his assistants dug through the gravel and the muck, ignoring the overpowering stench of rotting bodies in the grave. "I was amazed that those young men could handle the fragrance in the grave," Hultin said. "I was an old pathologist, I was used to it." The next day, their luck changed.

It was in the waning hours of the afternoon when Hultin saw it, in a hole seven feet deep; the body of a woman who had been about thirty years old. On either side of her were skeletons and badly decomposed corpses, but her body, though decaying, was mostly intact and her lungs were still well preserved and frozen. Hultin was awestruck. Why her? Why did she escape rotting down to mere bones?

"I sat on a pail—turned upside down—and looked at her," Hultin said. "Then I saw it. She was an obese woman; she had fat in her skin and around her organs and that served as a protection from the occasional short-term thawing of permafrost," Hultin explained. "Those on either side of her were not obese and they had decayed. I sat on the pail and saw this woman in a state of good preservation. And I knew that this was where the virus has got to come from, shedding light on the mysteries of 1918. I gave her a name. Lucy. Donald Johanson had sat in Ethiopia in 1974 and looked at a skeleton that shed light on human evolution. He had named her Lucy. I also thought of Lucy, *lux*, Latin for light. She would help Taubenberger shed light on that pandemic."

Carefully, he cut out both of the woman's lungs, placing them on a cutting board and slicing them, still frozen, with his autopsy knife. He placed them in a preservative that Taubenberger had provided him, a solution that would prevent the virus from deteriorating. He also took tissue from the lungs of three other bodies near Lucy that were badly decayed, although he had little hope that they would contain even fragments of virus.

Hultin knew he had to keep the tissues cold, but he felt it would be unacceptable to put his specimens in the refrigerator at the school where he was staying. Instead, he dug a tunnel in the grave, into the permafrost, and that became his refrigerator. It was a hole about two feet deep, which Hultin covered with boards. The next day, five days after he began, Hultin and his assistants began to close the gravesite.

Hultin had decided against looking for any more bodies. "We had found seven skeletons and three bodies with poor preservation, really poor," he said. "One, by chance, happened to be the body of an obese woman whose fat had served as a preservative. The likelihood of my finding anyone like that again would be very small. And I had good specimens—I could see it—well-preserved lungs." And if there was no virus in those lungs? Then, Hultin said, he would go back. But at the time, he thought he had succeeded beyond his wildest dreams.

"I was so privileged to have had the opportunity, so fortunate that Taubenberger did his work and that I was still able to go up there on my own," Hultin said.

After Hultin and the young men put back the earth and the sod that had covered the gravesite, Hultin had one more thing he wanted to do. When he had come to the gravesite in 1951, it had been marked by two large wooden crosses, but in 1997, the crosses were gone, the wood rotted. "When I met with the village council, I said, 'If I have time and if it is okay with you, I will leave only after I put two new crosses back for you,'" he said. "I got permission from the school to go into the woodshop and one evening I made two crosses."

Once the crosses were completed, Hultin set off for home. He had spent a total of $3,200, including his travel, and $900 in payments to his four assistants.

Taubenberger had contributed five dollars' worth of preservatives, a mixture of formaldehyde, ethanol, and guanidine.

Back in Washington, Taubenberger was waiting anxiously for word from Hultin. He was still dumbfounded by how well the project had proceeded.

Every night, Hultin would send him a fax: Sleeping on mattress in schoolhouse. A local man sold me salmon for $2 so I have no hunger. Got permission. Opened grave. Found skeletons. Found Lucy.

It had all happened so fast. "Basically, he wasted no time," Taubenberger said. He was struck by the contrast with Kirsty Duncan's group, which had spent six months just searching for the most experienced gravediggers and then signing them up. Hultin, Taubenberger said, "was there with a pickax. He dug a pit through solid ice in three days. This guy is unbelievable. It was just fantastic."

Hultin took his samples back to San Francisco, keeping them in an insulated and refrigerated pack. He worried about sending them to Taubenberger. If they got lost in transit, they could never be replaced. The only solution would be to divide them up and send them as four separate packages. He sent one package from San Francisco, using Federal Express. The next day, he sent another package, using United Parcel Service. On the third day, he sent a third package, using the U.S. Postal Service's Express

Mail. On the fourth day, he sent another package by Federal Express, this time from a town, Tracy, between San Francisco and Stockton, that he happened to be driving through. All arrived intact at Taubenberger's lab.

Hultin was almost finished with his part of the project. What remained was to memorialize the flu victims. "I located the office of the Seward Peninsula Lutheran Ministry in Nome and got the secretary there to correct the spellings and ages of the seventy-two flu victims and to thoroughly document their names and ages." He had two brass plaques made with the intention of attaching them to the larger wooden cross, which he did in September 1998.

Taubenberger, in the meantime, gingerly opened his precious parcels. Ann Reid began the lab work. Within a week they had it—the first solid evidence that there were fragments of the influenza virus's genes in the lung specimens from the woman that Hultin named Lucy. As Hultin suspected, no viral fragments remained in the badly decomposed tissue from the other flu victims.

Taubenberger called Hultin and told him the good news. And he asked Hultin how to proceed with the publicity, or if there was to be any publicity at all.

"We wanted to be clear to him, we wanted to be careful how we deal with the issue. The people of Brevig really were in control and we wanted to keep them informed. Hultin would do all the contacts." Through Hultin, Taubenberger asked the villagers how to proceed. "We asked them: What do you want to do? Do you want to have a press release? Wait for us to write a paper? Do something in between?" He told them if they had a press release they could be besieged by the world's media, pushing microphones in their faces. "Maybe you won't like that," Hultin cautioned.

From September until January, the Brevig residents thought about what to do. Hultin reminded them in telephone calls, in letters, and in faxes that he had to know their decision. In the meantime, Kirsty Duncan continued with her quest to exhume the Norwegian miners.

Duncan had no idea that Hultin had gone to Alaska and retrieved frozen lung tissue from a flu victim. Taubenberger hesitated to tell her, feeling bound by silence until he heard from the Brevig villagers. All she knew was that Taubenberger had one tissue sample, from the pathology warehouse, and was beginning to get the sequence of fragments of the virus's genes. Taubenberger and Reid, however, had gotten quite a bit more data. Their work was going quickly now. They had found traces of the 1918 virus in a second sample from the warehouse, in the lung cells from Private James Downs. They had determined the sequence of the hemagglutinin gene from the virus in Roscoe Vaughan's lungs, from the virus in James Downs's lungs, and from the lung tissue taken from Lucy, the Eskimo woman discovered by Johan Hultin. None of this was public, however, because Taubenberger was playing by the rules of science: don't announce your results unless and until they have been reviewed by a journal, accepted for publication, and actually appeared in print.

Unaware of the extent of Taubenberger's work, Duncan's group was continuing its elaborate efforts to be as certain as possible that it would not unleash a deadly flu virus upon the world. Lewin worked on designing special screws to remove tissue from the frozen corpses, for example. The idea was to use a method devised to take core samples of rings of trees, a drill that bores into the body, collecting tissue in a hollow tube. The group thought of just drilling into the corpses, but they were afraid that the drill itself might heat the tissue, possibly releasing a virus-containing aerosol. So they decided to simply twist the core biopsy screw in very slowly, by hand. They had been practicing on frozen pigs, perfecting their technique.

In October 1997, Duncan's group began its first investigation of the Spitsbergen graves, with ground-penetrating radar. The radar waves would pierce the ground in the Longyearbyen graveyard, bouncing off the earth and the coffins. The images do not show bodies or coffins, however; instead they show areas where the ground has been disturbed, by burials or coffins or even animals' digging. The art comes in interpreting the images to ascertain approximately the location of the miners' bodies, the depth of the permafrost layer, and whether the miners' bodies lay within the permanently frozen ground or in the so-called active layer that lies just above the true permafrost and that thaws and refreezes as the seasons pass.

"If they are really deep, like two to three meters," Lewin said at the time, "then we are convinced they will be in a frozen state and the information we could get with relation to viral extraction would be very great." The radar study would determine whether the group would return to Spitsbergen the next year or whether they would abandon the project.

In case the bodies were frozen, the scientists were already planning on protecting themselves and the world from a possible release of one of the mostly deadly viruses ever known. They would wear space suits, and the samples, Lewin said, would be placed in "containment units so there will be no chance of any germs being released during the exhumation."

"There is a very, very great danger," Lewin said. "In that frozen state there is a small possibility that the [virus] is still alive and there is a danger of [infection]. That's why we have to take great precautions in handling these bodies and the samples we will be getting from them."

The ground-penetrating radar study went ahead as planned, with the work being done by Sensors & Software, Inc., a Canadian company that has used its techniques to help police locate bodies in unmarked graves in the United States, Canada, England, and

India. Its system has also been used in archeology projects, including one to locate the "Lost Squadron" of fighter planes buried beneath the Greenland ice sheet. When the company directed its radar beams over the graves of the Norwegian miners, its pictures showed that all was well, the investigators reported. The radar "showed that there was a disturbance at a depth of about two meters," Duncan said, which, if the disturbance turned out to be the miners' bodies, would place them well within the permafrost layer.

It had been nearly two years since Duncan got permission to dig up the miners' graves and now, with the radar study, it was beginning to look like the project that had occupied so much of her time, energy, and emotions would actually go forward. Even the U.S. government was considering contributing funds. But first, Duncan and Dr. Robert Webster, the flu virologist from St. Jude Children's Research Hospital in Memphis, had to appear at a meeting on December 4, 1997, at the NIH to help answer a few remaining questions before the Institutes provided grant money.

Taubenberger was there that day although he had resigned from Duncan's group on September 7, having notified Duncan by fax of his decision. He explained to her that he could not help with her mission because he had heard from several reporters that she asked them to pay for interviews—a charge Duncan vehemently denies. Once the issue was raised, however, Taubenberger felt he had no choice. Charging the media for access to the Spitsbergen team "was incompatible with my position as a U.S. government scientist," he said.

The meeting room was studded with illustrious scientists—a famous virologist, a famous epidemiologist, a famous expert on respiratory diseases, a famous expert on emerging diseases. In addition to Kirsty Duncan, Robert Webster, and Jeffery Taubenberger, the participants were: Dr. Robert Couch, a microbiologist and influenza expert at Baylor College of Medicine, Dr. Nancy

Cox, chief of the influenza branch of the Centers for Disease Control and Prevention, Dr. Donald A. Henderson, a professor at the School of Hygiene and Public Health at Johns Hopkins University, Dr. Peter B. Jahrling, a scientific advisor to the U.S. Army Medical Research Institute of Infectious Diseases in Frederick, Maryland, Dr. William Jordan, an infectious disease expert at the National Institute of Allergy and Infectious Diseases, Dr. Edwin Kilbourne, the influenza expert from New York Medical College, Dr. Brian Mahy, the director of the Division of Viral Diseases at the Centers for Disease Control and Prevention, Dr. John LaMontagne, the director of the Division of Microbiology and Infectious Diseases at the National Institute of Allergy and Infectious Diseases, Dr. Pamela McInnes, the chief of the respiratory diseases branch of the National Institute of Allergy and Infectious Diseases, Dr. Timothy O'Leary, the chairman of cellular pathology at the Armed Forces Institute of Pathology, Ann Reid, who was Taubenberger's colleague, and Dr. John S. Spika, an epidemiologist at the Bureau of Communicable Diseases in Ottawa.

The agenda was chilling: "Biosecurity considerations," Robert Webster, 10 minutes. "Prevention of infection in exposed persons," Robert Couch, 15 minutes. Also on the schedule were "Prevention of spread off the island" and "Develop an algorithm for management of a disaster."

As the afternoon wore on, the group focused on what the Spitsbergen mission could do to advance scientific knowledge of the 1918 flu virus. It was the Taubenberger question come back to haunt them. What might they find that would justify opening those graves considering that Taubenberger was already getting viral genetic sequences from his material? and what, after all, was the risk of opening the miners' graves?

The Spitsbergen team argued that the sole source of information about the actual flu virus was a single case that Taubenberger had found, the tissue from Private Vaughan and the viral fragments in that tissue that had been the subject of his *Science* magazine paper the previous March.

Taubenberger was torn. The Brevig villagers had not yet decided if they wanted to go public. But clearly this group was misinformed. He decided to speak up.

"I said, no, we have three cases. The hemagglutinin gene sequence is finished and the sequences are identical," from the three specimens. "It went over like a bombshell. No one knew anything about this. The flu community thought the *Science* paper was my one shot. The fact that I had the complete hemagglutinin gene sequence and three cases threw them for a loop. There was just this kind of pause. Then I said that the three cases were from the fall wave of the epidemic. That was the important information. I just wanted them to know that, I thought it was important for them to know," Taubenberger said. But after the initial stunned silence, the group led by Kirsty Duncan went on talking about their plans, Taubenberger relates, "as though I hadn't said anything at all. They took this approach that we just didn't exist.

"It was the oddest meeting that you could ever imagine," Taubenberger remembered. "They wanted me to turn over the hemagglutinin sequence so that they could make a vaccine to immunize everybody in case there was a puff of virus released when they opened the grave."

Some suggested that maybe they should construct a specially made tent over the grave that would be a biohazard level 4 secure facility, essentially the most secure possible environment for working with deadly microorganisms like the Ebola virus or Lassa fever.

One man at the meeting was actually an expert on such facilities, Peter Jahrling, whose facility has one of the few such biocontainment labs in the world. Jahrling said that he was surprised when Robert Webster of Duncan's group first told him about the Spitsbergen project, but "it was actually very similar to something the Russians were doing—digging up bodies in the permafrost and looking for smallpox virus." That project began in the mid-1990s. "It sounded a bit less bizarre the second time around when I heard about it for flu." Although the influenza virus is much less stable than the smallpox virus, "I figured the prospects were probably not zero" that the group might succeed, Jahrling said.

But when Jahrling heard the group talk of building a secure biocontainment tent over the miners' grave, he could hardly believe his ears. Such facilities, called BL4, for biocontainment level 4, the highest level, are enormous undertakings. "I told them what engineering and operational controls go into a biocontainment facility," he said. "I showed them pictures of space suit labs, tight gas cabinets. We obviously can't duplicate that in the field. How much is reasonable to export to the frozen tundra?"

Jahrling explained that the investigators could wear hoods that filtered the air they would breathe and that they could wear disposable gowns and surgeons' gloves. Add to that "lots of bleach and you'll probably be okay," he said. But the group pressed him. Could they do more?

"At one point they were talking about whether they could approximate a BL4 tent. That was ludicrous. No, you can't."

On the other hand, no one present at the meeting thought that the chances of infection with the 1918 flu were very great. Some scientists guessed that there might be one chance in a billion billion that the miners' bodies would contain live virus, a number that still mystifies Duncan, who asks how anyone could be so precise.

Then Duncan's group presented the ground-penetrating-radar pictures. Taubenberger questioned them. The radar photos, he said, did not seem to him to show bodies beneath the permafrost. When he looked at them, he saw bodies that were close to the surface, in ground that had thawed. What was this project about? But Robert Webster said that only experts in ground-penetrating radar could interpret the photos. Unfortunately, none were present. Taubenberger, he noted, was no expert.

Only Edwin Kilbourne, the seasoned flu researcher, joined Taubenberger in skepticism that any viral genetic material would be recovered from the miners at Spitsbergen, let alone live virus. Kilbourne, like Taubenberger, was struck by the evidence that the tundra had frozen and thawed repeatedly over the years, making it almost inconceivable that viral genetic material remained. "I thought that if that was the case, the likelihood of preservation was small," Kilbourne said. He asked why the group needed powerful earth-moving equipment to uncover the bodies when they had been buried by gravediggers with pickaxes in 1918? If the ground was soft enough to be dug with a pickax in 1918, why did anyone think that the miners remained in frozen ground?

If the miners' bodies had thawed in the years since 1918, there was no hope of finding virus in them, Kilbourne told the group. He recalled the ubiquitous bacteria whose enzymes cause corpses to putrefy. Those enzymes also chew up viruses in body cells unless the tissue is kept frozen so the enzymes are inactive.

The more Kilbourne heard, the more he worried, but he asked himself if he was being too dismissive.

"At the conclusion of the meeting it was gradually revealed how much had already gone into planning," Kilbourne said, citing the ground-penetrating radar, the elaborate preparations for the expedition. He softened his harsh words. "I felt that if there was any chance of getting anything, go ahead. But it didn't seem to me to be a very well-thought-out proposal." After he returned home, Kilbourne wrote a long letter to the meeting organizer, Dr. John LaMontagne, "consolidating my position as being one of dissent."

Duncan and Webster got the federal money, a $150,000 grant. But Taubenberger left the meeting puzzled and angered by the reception he had received. Nancy Cox and Ed Kilbourne had paid attention to his announcement that he had a total of three samples with influenza virus. Duncan and Webster, he felt, had not. (Webster insists he never heard Taubenberger mention a third sample.) Taubenberger returned to his lab, across town in Washington, and went back to work. How long, he wondered, was he going to have to keep Hultin's work a secret?

In Hong Kong, the bird flu crisis was reaching its peak. It was nearly time to kill all the city's chickens, and scientists like Webster, Kilbourne, and Cox were acutely aware of how vulnerable the world was to a killer strain of flu.

The next month, as Duncan was about to set off for a meeting with her group in London, Taubenberger decided that he had to tell her about Hultin's expedition to Alaska and the tissue samples he had brought back. Duncan was stunned by the conversation.

"Jeff called on Friday at three in the afternoon. He started the conversation by saying, 'Kirsty, I hope you don't think any malice was intended but I have something to tell you.'" He proceeded to explain what Hultin had done.

"It was devastating," Duncan said. She felt betrayed by the secrecy. "I really felt that we were friends. I've never experienced anything like it."

Taubenberger felt that he was being as honest and straightforward as he could in a delicate situation and that he was making a real effort to be sure that Duncan had an opportunity to rethink her expedition. One reason for going ahead with the Spitsbergen mission was that Taubenberger's viral fragments had been soaked in formaldehyde for decades—the virus might have been altered by its long chemical bath. What was needed, they said, was virus that came from frozen tissues. Now, with Hultin's work, Taubenberger was telling Duncan that his group had it, the long-sought frozen lung tissue from 1918. He asked her to arrange a conference call when she was in London so that he could discuss Hultin's work with the entire group. Duncan called Taubenberger from London and told him that it would be impossible. It was his impression that Duncan was telling him that it was technologically impossible to have a conference call. Duncan, however, said that actually she had asked her team if they wanted to speak to Taubenberger and they had said no.

Five days before the London meeting, by coincidence, the residents of Brevig reached a decision. They would issue a press release. Taubenberger's group decided to follow it with a press release of its own.

Before sending it out, Taubenberger called Nancy Cox of the Centers for Disease Control and Prevention, Ed Kilbourne, and Dr. Dominick Iacuzio of the National Institutes of Health, wanting to make sure that the leading flu research groups knew in advance about the press release and what it would say. Duncan, of course, also knew; Taubenberger faxed the release to her.

Meanwhile, in London, Duncan's group decided to move forward with the journey to Spitsbergen. The stunning news of Hultin's expedition, however, did take them aback.

Not that John Oxford, the virologist from the Royal London Hospital who was a member of Duncan's group, would have voted to stop the expedition. Members of the group expected to get tissue samples from flu victims in another part of the world; they did not plan to preserve the samples as Hultin did, but to work directly with the fresh tissue, and they expected to get tissue samples from other organs as well as the lungs. "We thought we would gather more information," Oxford said.

Duncan left for Spitsbergen on August 14, 1998. It was time for the moment of truth after five years of planning and anticipation. As always in science, results can be a matter of luck. Yes, Hultin had found Lucy. But he might just as well have found skeletons.

And Duncan might find nothing but skulls and bones or she might find perfectly preserved bodies of the young miners. It would be chance either way.

Duncan reminded herself of her guiding principle: You don't know what you have until you actually do the work. "Even though all the evidence, historical, archeological, ecclesiastical, and radar, suggests there was something there, I was perfectly prepared that there might be nothing. My family is like, Are you excited? I told them no, it felt like I was coming to the end of a journey," Duncan said. "From a human point of view it was so hard, it was just so hard."

No detail had been spared in the planning and in assurances of safety. The group decided that at every stage, extreme caution should prevail. For example, even when the investigators were drilling into the frozen bodies to remove tissue, they should drill slowly so there would be no aerosols. And they should wear specially protective suits.

Duncan also insisted on treating the dead with honor and respect and reverence. When they arrived, the team of investigators gathered at the Spitsbergen cemetery and bowed their heads in a brief moment of silence to remember the young miners who were buried there, and "to give thanks to the young men's families and to the people of Norway, to give thanks for a great gift," Duncan said.

But it was hardly a private moment. The world's press was there, invited by Duncan—documentary crews, reporters, cameras for the television show *Nova*, recording every tear, every bowed head, every hushed moment. They were to remain at Spitsbergen for the entire exhumation, kept away from the graves by edict from Duncan but ready to interview anyone and everyone from the team and to photograph the expedition day by day.

Esther Oxford also was there, casting her sharp eye on the proceedings.

"We had the pathologist in the graveyard thanking God 'for the majesty of his creation,' and then posing for the cameras in a biological safety suit complete with respirator. We had a microbiologist trying to lecture us on virology. Most entertaining was Dr. Duncan [or Professor Duncan, as she was now referred to.] For five long days we had Kirsty Duncan talking endlessly about her hurts, hopes, and fears at the graveyard. Kirsty Duncan wearing short-skirts/latex leggings/sexy suede and high heels at the graveyard. Kirsty Duncan laying a wreath and demanding a moment's silence—at the graveyard. 'What are we going to have next?' joked a cameraman. 'Kirsty Duncan carrying the Olympic torch?'"

And yet, for all of her cynicism about Kirsty Duncan, Esther Oxford confessed that "I quite liked her," admiring her passion in the face of scientists who were so practiced at seeming neutral and objective. Before long, Esther Oxford wrote, "I was bewitched."

The exhumation started off with a meticulous preparation of the site. The group erected perimeter fencing to keep the site as private as possible. When that was done, they gently lifted the crosses from the graves, wrapped them for protection, and put them in a storage area. Next, they laid down blue matting material to protect the fragile Arctic ground. Finally, they began hauling their seventeen tons of supplies up the mountainside. They took rolls of film to record the position of every rock, every curbstone, so that when the excavation was completed, the group could return the site to its original condition.

The conditions of their permission to dig were that they bring in no wheeled or tracked vehicles. The sixteen men on the team and Duncan began hoisting the equipment with winches and by hand. Two tents were carried to the gravesite, one weighing half a ton and the other a quarter of a ton. It took ten people five hours to complete the task. The tools for excavating were carried by hand to the gravesite, which took almost four hours. Then a large bright blue tent was placed over the gravesite, ready to be inflated the next day; the group also installed a chemical decontamination shower in it. They unpacked their medical supplies, hoisted a backup propane-powered generator into place. The autopsy team practiced putting on and taking off protective clothing—space suits with filtered air supplies and puncture-resistant gloves, arm sleeve protectors, full aprons, knee protectors, thermal socks and boots.

On the fourth day, the group worked in a bone-chilling rain, completing its construction of protective walkways into the site. They built a storage area for the turf and curbstones they removed, and they moved lumber, plywood, and tools to their work area. Late in the afternoon, Pastor Jan Hoifodt of the town's Lutheran church led the group in prayer, then went into the tent and lifted out a portion of the sod covering the miners' graves. It was time for the Necropolis Company of London to move in, carefully removing pieces of tundra from the surface of the gravesite and storing them so they could be put back in their original positions.

The next day was a Sunday, and many members of the group began the day by attending church; work did not begin until the afternoon. While the rain continued, the team carried a freezer up the hill and put it under the tent. They prepared electric lines and nearly completed constructing their storage area for the tundra that was being removed from the surface of the graves.

The next day, the team found the coffins.

It happened so fast. Instead of having to dig deep into the earth, they came upon the simple wooden boxes just beneath the surface, in the active layer of the permafrost. Everyone was taken aback and Duncan became the target of a media swarm. What did this mean? Was the mission a disaster? Was it all a media extravaganza with no payoff? How could this be when the radar had indicated that the coffins would be well below the active layer, which gave the group a chance, at least, of finding well-preserved frozen bodies? Lewin himself had stated a year before that if the bodies were in the active layer, the chance that they would be badly decomposed was "very, very great."

Duncan responded with a statement to the press that gave no hint of despair. "The team of scientists and workers are extremely excited by these findings," she reported. In an interview seven months later, she emphasized her refusal to consider the mission successful only if the bodies were intact and frozen. "I was prepared that there could be nothing," Duncan said.

As a restive press corps gathered, including ten teams of documentary filmmakers, Duncan told her story through press releases and endless rounds of telephone interviews. She would awaken at a quarter of seven each morning and write a press release for the 9 a.m. press conference. After working at the site all day, she would return to a stack of pink telephone message slips from reporters. Once back in her room, she would start to call the press, finally taking her phone off the hook because the media would keep phoning through the night if she did not.

She made her private peace with the miners, going into the tent after the coffins were opened. Originally, she was not going to be present. Only a pathologist, his assistant, and someone to take samples were to be there, in order to minimize the chances of a deadly virus spreading. They had planned to take an experimental antiviral drug, a neuraminidase inhibitor made by Hoffmann–La Roche, to protect themselves, just in case. But now, with the coffins so close to the surface, the chances of a virus being present were essentially zero and there was no need to protect themselves. "I felt I had to be there," she said. It was silent in the tent, reverent. Duncan knelt beside the graves and thanked the miners.

The pathologist removed soft tissue from the miners' bodies, Duncan said. But she declines to say what sort of tissue was obtained. "I cannot release that information," she maintained. Family members "have given me this great gift. If those were my grandparents, I would not want someone talking about what they looked like." She said in March 1999 that the laboratory analysis of the tissues had just begun and so it was too soon to say if any viral material was there. But, she adds, she is proud of the mission because it showed that such an effort can proceed safely and ethically.

Yes, Duncan said, it cost half a million dollars, most of which came from grants. But, she insisted, "safety costs."

Lewin, for one, claims that the project was a great success.

"We felt we had only a very short time span to go up there," he said. "It worked like clockwork, it was a superb project, a template for any future project." He is particularly proud of the care and respect with which the team operated. "The most important thing was that all the precautions we took were such that the local community was very grateful," Lewin said. "Not only did we take care of the surroundings but we made sure that there was no danger whatsoever to the community. We had no idea what we would find. To me, it was very exciting."

But what exactly did they find? The problem, Lewin said, was that "it was one of the warmest summers ever, the whole area had thawed out. The bodies in their caskets had been squeezed out of the permafrost. They were quite putrefied, those bodies."

Nonetheless, the group did get samples of soft tissue. "We had over a hundred samples, lots of soft tissue, a lot of stuff. It is now being analyzed; it is a very complicated investigation." Asked what sorts of soft tissues were obtained, Lewin replied that it was mostly brain, but also included muscle, lung, "all the tissues you can think of."

"This material wasn't like a fresh body you see in a morgue. It was badly decomposed," Lewin explained. The bodies, he adds, "weren't frozen. For at least six months or a year for the last two or three years they were probably thawed and frozen, thawed and frozen."

But is there flu virus in the tissue? Lewin won't say. "We decided, no sort of advance notice will be given unless we have something definitive to report." And the group is not ready to share its samples with other scientists. "We did not want to distribute tissue until we are satisfied that there is no danger of infectivity," he said.

John Oxford had a different view of what had been accomplished.

"We were terribly disappointed, it has to be said. The virologists were frightfully disappointed. In my expectations, we would find seven perfectly preserved young coal miners. What we found, more or less, were seven skeletons with tissue on them. We ended up with a couple of dozen biological samples," which were sent to a British government laboratory that has the biocontainment facilities to study deadly viruses like Ebola.

But Oxford is not very hopeful.

"The brain samples are the best we've got. And when you think about it, the chance of getting virus information out of six brains when you're not sure that the brain had anything to do with it is a bit remote."

Duncan has not given up. "Science is about trying," she said. "Sometimes you get an answer and sometimes you don't. At this point, we don't know."

But she is left with a bitter taste about the scientists who at first embraced her, and then turned against her, and who almost to a person had different versions of events than the ones that Duncan remembered.

"Everyone had their own agenda," Duncan said. "For most of them, they're flu experts; they had a lot to gain or lose. All I wanted was an answer.

"I take the high road. It's not my nature to say anything, but there's only so much one person can take, there really is. I've been lied to by many, many people, from government agencies to individual scientists. It's been the most unpleasant experience of my life."

# 10

# MYSTERIES AND HYPOTHESES

At the turn of the millennium, scientists were left with two mysteries and a handful of hypotheses about the 1918 flu.

The first mystery was: Where did that flu come from? It had appeared seemingly out of nowhere, and had killed people all over the world. There was no obvious source for the deadly flu strain and the stories and myths that had cropped up to explain the flu seemed preposterous to most flu experts. The most widespread explanation, and the one advanced by a television documentary, *Influenza 1918*, which aired on "The American Experience" in 1998, suggested that the flu had arisen at Fort Riley in Kansas, where soldiers were housed in a farm region, not far from pigs. It was the burning of the pig manure, huge black clouds of it, that spread the flu, according to this notion.

Flu experts scoff. "Starting from burning manure is manure," Jeffery Taubenberger shot back.

The flu virus could not possibly have been spread this way, flu experts maintained. It is a fragile virus that dies on its own when it is outside the body. There is no reason to believe that pigs gave it to humans—it is at least as likely that humans gave it to pigs. And the evidence suggests that it is at least as likely that the deadly flu strain came to America from Europe as the other way around.

How, then, to solve the mystery?

John Oxford will never forget the moment he came upon what he thinks is a solution. It was late August 1998, and Oxford, a member of Kirsty Duncan's group, had arrived in Spitsbergen, to be present when the miners' graves were opened. He was there more to observe than to participate; his actual duties at the gravesite were minimal. Oxford spoke to the hordes of reporters, the omnipresent television crews, and the swarms of documentary filmmakers, radio broadcasters, and the world's print media that had invaded the lonely Norwegian island. But eventually he found himself moving away from the crowd, mulling over the mission and the quest to understand the 1918 flu.

The story of the stricken young men who had died before they had a chance to begin their work in the Spitsbergen coal mines haunted Oxford. He could imagine the dank cold and the gloomy days when the seven men, already hot with fever, stumbled off the ship that carried them to Spitsbergen from Norway. He could imagine their labored breathing as the influenza virus took hold and their lungs filled with fluid. And he could imagine their agonized deaths and solemn burial in the frozen ground of the cemetery on the side of the lonely mountain a few days later.

Oxford, an elfin-faced man with a pensive bent, could not get the thoughts out of his mind. What would it be like if that flu virus returned? How would you know it had arrived, creeping silently in a small town in Europe, in a teeming city in China, or spreading in an airplane cabin as a passenger sneezed and coughed? How could you find it in time to stop it?

Oxford recalled the close call in Hong Kong that past winter when the deadly bird flu emerged. He blanched at the thought of how close the world had come to another terrible pandemic. For all the advances in molecular biology and genetics in the years since 1918, humanity still seemed at the mercy of a genetic twist that could create a killer virus out of an ordinary flu.

As a cold rain began falling in Spitsbergen, Oxford took refuge in a hut at the bottom of the mountain overlooking the gravesite, drinking steaming cups of strong tea, reading books about 1918, and thinking about flu epidemics and their victims. He recalled a poem about the Spitsbergen miners that he had written, titled "He Won, Didn't He?" It read, in part:

*I can picture you now on your journey,*
*Thoughtful, enterprising, tough.*
*You were the miners.*
*You would excavate in that permanently frozen land.*
*You had all the equipment and you said your goodbyes as you left the Hanseatic port in 1918.*
*But our mysterious infective friend traveled with you, in you and, in the end, he won didn't he?*
*At first when you felt the ache, you passed it off—seasickness, anything but that.*
*In the end, though he won, didn't he?*
*You were shocked when your first friend died, and then the next.*
*In the end seven of you were laid in the small village hall with the flickering candles, the ever-present snow and wind.*
*So they chose a spot on the edge of a cemetery deep in the ice where together you would lie until the end of time, perfectly preserved, fixed, all tissues intact and frozen forever.*
*So how will you view us, as serious explorers after truth?*
*Will you be pleased or saddened?*
*Will fear strike you, and will you resist or will a glow of pride pass through you?*
*I think you will be pleased to help.*
*It will be painless. We promise.*
*Just a brief exposure to light as we uncover you.*
*Not completely, you understand, we want you to remain frozen forever. Just as you are now.*
*You will not see me, but just my friend in white, masked like a surgeon. But don't worry.*

*Just a small prick as we take our lung sample, like a biopsy, really.*
*You won't mind?*
*With your help he may not win in the end.*

As the days passed and it became clear that Duncan's mission had not succeeded, Oxford realized that the only hope of finding the 1918 virus lay with Jeffery Taubenberger, working in his lab on the tissue from the warehouse and from Alaska. He mulled the twist of fate that had put the Norwegian miners' bodies so close to the surface, writing another poem that he called "Hurried Burial 1918." It read in part:

*So in the last analysis your burial was hurried.*
*The dynamite blasted the hole but in darkness at –20 C your friends did not excavate deeply.*
*They placed the seven of you tightly together, side by side for company.*
*But one of them had walked to the beach and that bucket of sand had served as the "earth to earth to dust" message.*
*It was the first thing we found eighty years after as we searched your grave.*
*Year by year you had forced your way skywards, hoping, expecting to see day.*
*With our help you did, but as much as we looked we could not reconstruct you.*
*A skeleton of a healthy young man was the pathologist's reports.*
*Well, you were not that healthy . . .*
*But now it's over.*
*No disturbances until the end of time, to resurrection.*

Oxford sat in the hut on the mountainside, thinking about the victims of the 1918 flu, when all at once, he said, an idea "came at me like a rocket."

"There were all these young people, all unbeknownst to each other, on different continents. And a virus was striking them," Oxford realized. How could the same virus, that could survive for at most a few hours outside of a human body, that was spread by direct contact, through droplets of saliva or mucus, end up infecting people all over the world on virtually the same day? Perhaps, Oxford thought, the virus was already present, smoldering in towns and cities throughout the world. Perhaps, he thought, the 1918 influenza virus had arisen before 1918. It was a chicken-and-egg sort of question. Clearly, the virus had to come from somewhere. Clearly, it had to arise sometime. If it did not arise in 1918, then when?

Oxford is a traditional virologist who has yearned to understand the 1918 flu for all of his professional life. He has earned a reputation as a sort of British counterpart to Jeffery Taubenberger and Johan Hultin, one of the few virologists in the world who were seriously looking for lung tissue from flu victims who died in 1918. But unlike Taubenberger and Hultin, Oxford had studied virology all along and had known about the 1918 flu from his earliest days as a scientist.

Even the year 1918 resonates for Oxford. He was born much later—in 1942—but his father had been an airman in World War I and returned from fighting in France in 1918. Almost every year, he would take young John to the commemoration of the war's end on the eleventh hour of the eleventh day of November. In later years, the Queen would attend the ceremony in London with the royal family. Big Ben would chime. Then all who assembled would bow their heads for two minutes of hushed silence to remember those who died in the Great War and the peace that had finally arrived at 11 p.m. on November 11, 1918.

"It was an evocative time, 1918," Oxford said. And for him, a large part of the reverence attending that year was the flu, a disease, he said, that "rose up from the ashes" of the war.

All virologists of his generation knew about the 1918 flu, Oxford said. "After all, that's the biggest outbreak of any infectious disease ever known, bigger than the Black Death." He estimates the number of deaths worldwide as 100 million, a larger number than the conventional estimate of 20 to 40 million. But, he said, 20 million people died in India alone, making it impossible for the 20 to 40 million figure to be correct. With so many dead, that flu epidemic profoundly affected the lives of ordinary people whose voices are seldom heard.

One woman sent Oxford a photograph of her parents, Thomas and Gladys Frederich, on their wedding day. On September 7, 1918, within months of their marriage, Thomas caught the flu and died, leaving Gladys alone and forced to fend for herself. When she died, decades later, that wedding picture was still at her bedside.

"One person dies and there are repercussions through the next generation. Multiply that by 100 million and then you begin to see the effects of the flu," Oxford said. The photo haunts Oxford, reminding him of his mission as a scientist—to figure out where that virus came from and how to stop it if it returns. He had spent much of his life steeped in flu virology, with but one, ultimately unsuccessful, foray into another arena.

Oxford had begun his education in virology at the University of Sheffield under the tutelage of Sir Charles Stuart-Harris, a member of the team that isolated the first human influenza virus in London in 1933. Stuart-Harris was the one who had had to check the ferrets each day after the British team discovered that those fierce rodents were almost uniquely susceptible to human influenza viruses, developing all the symptoms that plague people: fevers and drippy noses, muscle aches and pains, and general malaise.

Oxford knew this about Stuart-Harris, but, being young—he was Stuart-Harris's student when he was nineteen and twenty years old—he did not bother to quiz him about the details of his experiences trying to unravel the secrets of the flu of 1918. Now it is too late. Stuart-Harris died in 1997. "There are millions of things I would have liked to ask him, but he's gone," Oxford said.

Instead, when he worked with Stuart-Harris, Oxford concentrated on the influenza epidemic of 1968. He was part of a team that was scurrying to develop vaccines and what they hoped would be flu-fighting drugs while the illness burned through the world's population.

"It was a firefighting exercise," Oxford said. That flu epidemic ended, of course, and after a lull HIV appeared, making influenza seem like yesterday's problem.

"Influenza began to seem less and less exciting," Oxford said. The flu virus had been discovered years ago with the work of Professor Wilson Smith, Sir Christopher H. Andrewes, and Sir P. P. Laidlaw in 1933, when they transmitted human influenza to ferrets. Everyone knew what its genes were—they had been found in 1968. Flu did not seem so terribly dangerous anymore. And there were fewer and fewer epidemics.

"Lots of influenza virologists, including myself, became diverted to HIV," Oxford observed. HIV was a virus that scientists had never seen before and no one knew what course it would take as it spread through the world's populations. As young men sickened, suffered terribly, and died, there was a scientific and moral imperative to understand the virus and find a way to stop it. As the virus began to decimate African nations and to destroy a generation in Europe and America, as homosexuals, intravenous drug users, and their sexual partners were felled, increasingly many doctors and virologists took on the study of HIV as a personal mission, applying for the suddenly abundant research funds, turning over their laboratories to the study of the virus, embarking on a quest to vanquish this terrible killer.

John Oxford was among these converts, one of many who confidently assumed that with today's knowledge of molecular biology and tools for discovering medicines and vaccines, they could figure out HIV and save millions of lives.

"Every virologist loves a new virus," Oxford remarked. "Influenza virologists thought, 'Here's one that can be knocked off like pins in a bowling alley. Here's an easy one.'" Of course, they were wrong.

For Oxford, the flu began to seem interesting again.

He turned his attention back to the 1918 flu, although he knew he was going down a lonely path, with no obvious guideposts along the way. Virologists worked either on HIV or on what Oxford calls "hospital viruses," ones that, like herpes or hepatitis, send patients to the hospital. Garden-variety influenza, Oxford asserted, "is not a hospital virus."

But, Oxford realized, the challenge would be to understand what might turn ordinary influenza into a hospital virus, which meant figuring out what the 1918 flu looked like. One day, a colleague, Dr. Rodney Daniels, reminded Oxford of a unique repository at the Royal London Hospital, where they worked, that had been all but ignored over the years. Deep in the bowels of the pathology building was a vast collection of clinical samples that dated back to 1900. Daniels turned to Oxford and said, "Why don't you take a look at those old samples in the collection. There might be some interesting flu stuff. You shouldn't miss something on your own doorstep."

There, in a hot, dark room, with mobile stacks that run along rails, are wooden boxes that look like orange crates. In those boxes are hundreds of thousands of cardboard boxes with patients' numbers on them. And in those boxes are preserved snippets of tissue from patients who had been admitted to the hospital. The collection included pathology specimens from people who died in 1918, making it a small-scale British equivalent of the vast pathology warehouse that Taubenberger had used to find the lung samples from soldiers who perished from the 1918 flu. Without knowing about Taubenberger's work, Oxford began pursuing the same idea.

The samples from 1918 were collected by Dr. H. M. Turnbull, who worked in the hospital during World War I and watched, helplessly, as young soldiers fell ill and succumbed to the flu.

Unable to figure out why the disease was so deadly, he gathered as many samples as he could from the lungs and brains of soldiers at autopsy and preserved them in wax. It was as though science were a relay race and the pathologists of 1918 were the equivalent of the first runners in the race, whose job it was to pass on the baton, Oxford observed, saving samples for future generations to study.

Inspired, Oxford began his search for tissue from flu victims. "There were huge books, like Bibles, from 1918, with postmortem numbers in them," he said, identifying the patients whose tissue was stored in the basement. Along with a postmortem number came an elaborate pathologist's report.

"Everyone who dies has a postmortem telling when they died, their occupation, their age, their diagnosis at the time. The pathologist would take out the vital organs, fix them and section them and look at them under a microscope. He would try to do a diagnosis. He would write it all in a book, with microscopic handwriting," Oxford said. The task was arduous. "It didn't take a rocket scientist to appreciate that a lot of young people were dying of pneumonia. I had to go through the postmortem book and pick out every sample of someone who died of pneumonia and note the postmortem number. Then I had to whisk down to the room," he said, and search for the samples.

For the past few years Oxford has employed a group of students who spend their summer vacations rummaging around in the hospital basement, searching for tissue from victims of the 1918 flu. When Taubenberger wanted to find promising samples, he did a computer search of the warehouse's holdings, then put in his order. The samples arrived on his desk in a few days. When Oxford's assistants wanted to find samples, they peered through the old books, went to the hospital basement, and then began what could be a long and tedious search for the box that held those samples. Then they had to carefully check to be sure the box's labels were correct.

Eventually, the students found eight samples, snippets of lung tissue soaked in formaldehyde and encased in small blocks of wax that had come from patients who died of the flu in 1918. They had no idea that Taubenberger was doing the same thing. But unlike Taubenberger's samples, theirs seemed no longer to contain shards of the flu virus.

It was this work with the pathology samples in the basement of his hospital that made Oxford one of the few experts—or, at least, would-be experts—on the 1918 flu. Duncan took note and telephoned Oxford the year before her group went to Spitsbergen. He invited her to come to London so they could talk at length about her project. The two hit it off immediately, Oxford said, and he eagerly agreed to be part of her expedition. He also agreed that it was important to be extremely cautious.

"I certainly felt that if these seven men were going to be found in good condition, then we would have to be very careful. Given that this virus killed 100 million people we did not want to be responsible for another outbreak," he said. "We did a safety analysis, a very careful one. Although we thought it was very unlikely that anyone would get infected—how could you get infected from a frozen corpse?—scientifically we thought it was important that we took all the precautions we could think of."

The precautions meant that when the group finally arrived in Spitsbergen, the exhumation was slow and tedious, giving Oxford plenty of time to sip his tea and think. It was while the graves were being opened, while he was sitting in the hut, writing his poetry, and the idea came to him that perhaps the 1918 flu—the Spanish Lady—had already seeded itself around the world by 1918, that his mind leapt to the next steps. First, he needed to look through the medical literature to see if there was any evidence that people had been infected with the virus before 1918. And then, if he found such evidence, he had to search for that virus in the lungs of people who died of the flu before 1918. "I am not even talking about the Spanish Lady of 1918," Oxford said. "I am talking about a Spanish Lady from 1916 or 1917."

Oxford went home and rushed to the library, where he scoured old journal articles for case histories of people who became ill with a 1918-like virus before 1918. It did not take him long to find what he was seeking. In 1916 and 1917, he discovered, doctors were reporting outbreaks of what looked like a lethal respiratory virus among troops in a British camp, Aldershot Barracks, just outside of London, and in British Army camps in France. The disease was called catarrh, not influenza, but its symptoms were eerily like those of the 1918 flu. Victims would develop cyanosis—their ears and lips would turn blue from a lack of oxygen—and large numbers would die. Oxford points out that when influenza patients were admitted to hospital wards in 1918, nurses and doctors said they could pick out the ones who would die by looking for cyanosis of the ears and lips.

One report was published in the *British Medical Journal* on July 14, 1917. "Purulent Bronchitis," it was called, subtitled "A Study of Cases Occurring Amongst the British Troops at a Base in France." The authors, medical officers in the British Army, reported on a disease with "features of clinical and pathological interest."

The sickness first appeared in the camp in December 1916; a month later it became "a small epidemic," they wrote. Its symptoms were hauntingly familiar to Oxford—the British doctors could have been describing the 1918 flu: A group of patients shows up at the hospital with a temperature of about 103 degrees and coughing up blood-streaked pus. The patients have a rapid pulse and soon show signs of cyanosis. They die from asphyxiation as their lungs fill with fluid. Others have a less severe disease and recover, but only after several weeks of fevers and wasting.

Less scientific but hard to forget are personal stories that Oxford began collecting. One woman, for example, told him that her father was in Toronto when World War I started but volunteered to fight and came to England in 1915. Her father had said repeatedly that he saw a frightening disease in the camps. Many of the men, he said, contracted influenza, and many died of it. But, with the wartime secrecy imposed, the soldiers kept mum about the sickness and deaths.

The more Oxford thought about it, the more it seemed that an earlier date of onset for the 1918 flu made perfect sense. It fit well with another medical mystery, a terrifying epidemic of a new brain disease that appeared in Europe and North America between 1916 and 1926. The disease, a kind of sleeping sickness known as encephalitis lethargica, killed an estimated five million people before it abruptly disappeared. Some have argued that it was a consequence of the flu, but if so, how could it have begun before the 1918 epidemic? Unless, of course, the epidemic began before 1918.

The sleeping sickness was first described by a Viennese doctor, Baron Constantin von Economo, in a paper published in 1917. "Since Christmas," von Economo wrote, "we have had the opportunity to observe a series of cases at the psychiatric clinic that do not fit any of our usual diagnoses. Nevertheless, they show a similarity in type of onset and symptomatology that forces one to group them into one clinical picture. We are dealing with a sleeping sickness, so to speak, having an unusually prolonged course."

Patients would sleep around the clock, and although they could be roused and could answer questions and follow commands, it was as though they were sleepwalking.

"If left by himself, he falls back to his somnolent state," von Economo wrote of these patients. Some died within a few weeks, others lingered for weeks or months, falling into periods of deep sleep punctuated by comas. Those with this prolonged course of illness sometimes survived, but if they did they never fully recovered. When the crisis was over and they were no longer ill, they would sit motionless, aware of their surroundings, but lethargic and unresponsive, like extinct volcanoes, von Economo said.

Many developed a form of Parkinson's disease, a neurological disorder that occurs when part of the brain that controls movement is destroyed. They would remain frozen, unable to move or respond, their thoughts and emotions inscrutable behind their masklike faces.

In investigating what caused the illness, von Economo searched for microorganisms in patients' brains that had been removed at autopsy. The brains contained a virus, he discovered, that could transmit the disease to monkeys. But he was unable to isolate the virus or determine what sort of virus it was.

Von Economo noticed that some sufferers initially became ill with a respiratory disease. But many did not. He specifically stated that he did not think the new illness was a consequence of the "grippe" that was spreading through Europe. But other medical scientists in years since have argued that the sleeping sickness was a peculiar result of the 1918 flu.

In 1982, R. T. Ravenholt and William H. Foege, two scientists at the Centers for Disease Control, made this case based on epidemiological data from Seattle, Washington, and the Samoa Islands.

In Seattle, the scientists reported that there appeared to be a direct connection between influenza and the deaths of 142 people who died of encephalitis lethargica. Local newspaper reports tended to connect the diseases with influenza. One article, in the November 29, 1919, issue of the *Seattle Times*, described two cases of the disease. "This is the first appearance of the disease so prevalent now in England and reported from various isolated parts of the United States. While it has not been definitely determined that the disease is an aftermath of the Spanish influenza, both of the Riverton persons were influenza victims last year," the article said, referring to inhabitants of a Seattle suburb.

In addition to the 142 people who had the flu just before they developed encephalitis lethargica, there were 18 other deaths from encephalitis lethargica in which the time of onset was not well established. Nonetheless, sometimes there were intriguing connections. A sixty-one-year-old woman who died in 1924, for example, was described in her obituary as dying from "flu followed by paralysis during the six years from 1918 to 1924."

Most scientists, hearing such evidence, would question it. It seems to fall into an old logical trap—an assumption that because one event preceded another, the first event caused the second one. It was the sort of snare that caught those who argued that the swine flu vaccine caused medical problems. If you immunize millions of people, then of course, by chance alone, some will die soon after they are injected with the vaccine, others will have strokes or heart attacks. That does not mean that the vaccine caused the deaths, strokes, or heart attacks. Most people alive in 1918 got the flu. Even if encephalitis lethargica had nothing to do with the flu, by chance alone most people who got encephalitis lethargica would also have had the flu. To make the argument stick, the scientists needed more than just a correlation with influenza in a population that was hard hit by the flu. They needed a situation that made the case.

It turned out that there was one place in the world where there had been an inadvertent experiment that might be just what was needed. It took place on the islands of Samoa.

The flu came to the islands of Western Samoa on November 7, 1918, arriving with the crew of a steamer, *Talune*, from Auckland that had docked in the tropical paradise. In the next two months, 8,000 people on the islands died of influenza, a fifth of the population.

The experience in Western Samoa was a blaring warning to the inhabitants of American Samoa, a little more than a hundred miles away. The residents were terrified that the disease would arrive there, and to keep it out, they barred all contact with the outside world. With a strict quarantine, the American Samoan islands managed to escape the flu.

Now, the question is: What happened with encephalitis lethargica? Did it occur regardless of the 1918 flu? The data support the hypothesis that the flu strain set off the brain disease in susceptible people. In Western Samoa, seventy-nine people died of encephalitis lethargica in the years 1919 to 1922. In American Samoa, two people died of the disease.

The evidence is provocative, not definitive, Oxford understands. But, he said, it is nonetheless hard to dismiss, which brings him back to the pathology samples stored in his hospital. They might provide an answer. Was the 1918 flu strain present earlier? Did it attack the brain?

Oxford sent his students back to the basement, looking for lung tissue from people who died of the flu in 1916 and 1917. He found some promising samples, which he wants Jeffery Taubenberger to analyze as soon as he finishes his work with the samples from 1918. Taubenberger, in the meantime, also searched the database for the armed forces' warehouse, looking for samples of his own from 1916 and 1917. But he came up empty-handed.

While John Oxford's students were searching the dusty basement warehouse in London, Kennedy Shortridge in Hong Kong was also wondering whether the 1918 flu started before 1918. He had no stored sample, but he had another way to ask the question. Clues, he argued, were present in the molecular genetic research by Jeffery Taubenberger and Ann Reid.

Shortridge's interest was sparked by the near-miss in 1997, when the H5N1 bird flu virus infected people and killed them. Might the 1918 flu have done the same thing? The only saving grace in the 1997 bird flu episode was the quick action by Hong Kong to contain it. Yet, Shortridge wrote, "were it not for the slaughtering of 1.5 million chickens and other poultry," the world might have been hit by another pandemic.

The central fact about influenza pandemics, Shortridge emphasized, is that every one that has ever been traced to its origin began in China's Guangdong Province, formerly called Canton, which is in the south, next to Hong Kong. Even the very first recorded flu epidemic began there, Shortridge said, originating in Guangdong in September and October 1888. Could it be that the 1918 flu began as a bird flu in southern China?

The emerging work by Taubenberger and Reid certainly made it seem plausible. As soon as they realized that their molecular fishhooks would work to pull out fragments of the viral genes, the two had to choose which gene to concentrate on first. They chose the most obvious target, the hemagglutinin gene. That is the gene that provokes an immune response in infected people and it is the gene that virologists had suggested holds the clue to the lethal nature of the 1918 virus. It also can help answer the questions that Shortridge posed: Where did the virus come from? Did it originate in birds, like the H5N1 Hong Kong flu? Or did it jump from pigs to people, like the 1976 swine flu?

To answer that question Taubenberger's group had to compare the sequence of the hemagglutinin gene from the 1918 flu virus with those of hemagglutinin genes from flu viruses that infect only birds and others that infect pigs. Then a computer program decided the simplest genetic path to convert a bird virus's hemagglutinin gene to that of the 1918 flu and to convert a pig virus's gene to that of the 1918 flu. How many mutations would it take to accomplish the task? What is the smallest number of mutations necessary?

From the gene sequences, the computer program constructs theoretical family trees. "You can do these programs many different ways," Taubenberger said. There always are multiple paths leading to the same outcome. Yet no matter which path the group constructed, the bottom line was always the same: the 1918 flu resembled a bird flu but it could not have come directly from a bird—it had to have been adapted and modified first by growing in humans or pigs.

Shortridge suggests that the 1918 virus gradually changed over perhaps a fifty-year period from a bird flu to one that could infect humans. Eventually, it became a strain that would be deadly to virtually everyone in the world—everyone, that is, except people of southern China, who had been living with the strain for so long. In fact, he said, when he went back and looked at records from southern China in 1918, he discovered that the 1918 flu was not particularly deadly there, in direct contrast to its effects elsewhere.

"This virus had been kicking around for a long time," Shortridge said, "maybe co-circulating with other viruses."

If Shortridge is correct, his solution to the mystery of where the flu came from takes care of the question of how the virus got to Europe. Once again, Shortridge has an answer. During World War I, Chinese laborers traveled to camps in France and built trenches for the Allies. They could easily have taken the flu with them, sparking the epidemic that swept the world, he argues.

But not everyone is convinced by the evidence that Oxford and Shortridge marshaled.

Jeffery Taubenberger, for example, agrees that the 1918 flu virus may have begun circulating before 1918, citing his own evidence from the genetic sequence of the virus's hemagglutinin gene. By analyzing that sequence and asking how it might have arisen—by mutations from previous influenza viruses—Taubenberger and his group concluded that the 1918 virus might actually have emerged and started infecting humans sometime between 1900 and 1915. Moreover, he said, it is intriguing that the influenza death rates in the United States began rising in 1915 and continued to rise until 1917, when they dropped slightly, followed by a huge spike in influenza mortality rates in 1918. The question he cannot answer, however, is whether the gradual rise in flu deaths before 1918 was the beginning of the 1918 influenza pandemic or whether it was simply a reflection of modest changes in another, much less dangerous flu strain.

But Taubenberger is not won over by the papers that Oxford found on deaths in the British camps. "The 'purulent bronchitis' paper mentioned by John doesn't sound like flu and the description in the paper does not look like a good match to 1918 flu pathology," he said.

Taubenberger is also skeptical of Shortridge's hypothesis that the 1918 flu originated in China. "I have seen no evidence to support this." In fact, Taubenberger added, an "excellent paper" published in 1919, in English, in the *National Medical Journal of China*, shows that, at least in Harbin, China, the 1918 flu epidemic followed exactly the same pattern as it did in the United States and Europe: there was an initial wave of infections in the spring of 1918 with a flu virus that was highly contagious but not very lethal followed by a second wave in the fall with a virus that was deadly. That virus in the autumn of 1918 also infected pigs in China, killing them just as it killed pigs in the United States, Taubenberger noted.

"I conclude from this that by spring flu was already spread around the globe," Taubenberger said. There is, he added, "no more evidence that it started in China than in the United States. Either is still possible."

For Taubenberger, the mystery of the flu's origins and why the virus went unnoticed if it was around before 1918 remains very much unsolved. The crucial clue might lie in the preserved lung tissue from people who died of influenza before 1918, a treasure that, he hopes, he will be able to unearth from the Army warehouse.

• • •

A second mystery was: Why did the flu preferentially kill young people, between the ages of twenty and forty? It is a question that has troubled virologists because it is so contrary to the usual pattern of deaths from flu. Every other influenza virus has killed the very old and the very young, sparing healthy adults in the prime of life.

That mystery is easy to solve, asserts Peter Palese, the chairman of the microbiology department at Mount Sinai School of Medicine in New York.

Palese, middle-aged and deliberate, with pale blue eyes and wire-rimmed glasses, still has the formal demeanor of his native Austria. On a bright spring day, he sat at a table in his office, with his back to a large window overlooking New York's East River, and spoke of how he came to his convictions about the flu.

Palese has spent virtually his entire scientific career studying influenza and probing the virus's secrets. He started out to be a chemist, studying substances that could block neuraminidases, the very enzyme that flu viruses use to burst out of cells. That work led him to what became an abiding interest—figuring out influenza viruses and how they work.

By 1976, he had come to Mount Sinai School of Medicine, working under Ed Kilbourne, who was the chairman of the microbiology department. He watched with fascination as the elders of his field urged the nation to embark on a swine flu immunization campaign. Palese privately thought it was a mistake because, he believed, the Fort Dix virus was not a danger to the human population. It was a swine virus, he reasoned, and before a swine virus can spread among people it has to undergo genetic changes that would make it grow avidly in human lungs. Palese thought that the best course would be to make a swine flu vaccine and stockpile it in case the virus, against all odds, caused a deadly epidemic. He kept mum, however, because as a junior scientist he realized that he did not have the stature to advise the government on the swine flu.

"I was just a little assistant professor," Palese said. "There were not many people talking to me."

Instead, he concentrated on his work understanding the genetics and biochemistry of influenza viruses.

When Palese was asked why the 1918 flu killed young adults, he started by remarking on a well-known phenomenon: Virtually every viral disease is more deadly in teenagers than children, and more deadly in young adults than teenagers. Think of measles or chicken pox or smallpox, he said. They swept through Native American and Eskimo populations like wildfire, killing adults without mercy. But children who were infected had much milder courses. So the fact that the graph of deaths from the flu as a function of age is a straight line is no surprise. It would be expected that the older people became, the more deadly a new virus would be.

The question is why did the death rate drop off precipitously in people over age forty? The most likely explanation, Palese said, is that a similar virus, not as deadly, had come by earlier, providing some immunity to the 1918 flu virus to those who had been exposed to it.

Jeffery Taubenberger reached the same conclusion. But now he is left with the fundamental question, the heart of the murder mystery: What made the 1918 flu so deadly?

That led Taubenberger to the three hypotheses, each a plausible explanation for the flu's virulence. He attacked them one by one.

The first hope was that the virus's hemagglutinin gene would provide the clue. This, after all, was one of two proteins that poke from the surface of the flu virus. It is the protein that the virus uses to shove its way into cells, and when the immune system blocks a flu infection, one way it does so is to block the virus's hemagglutinin proteins.

The hemagglutinin protein is also the reason why flu viruses grow only in people's lungs. When flu viruses infect cells, they make a large precursor of their hemagglutinin protein that must be hacked in two by a cellular enzyme. Since that enzyme is present only in human lungs, the virus can grow only in lung cells.

One hypothesis about the 1918 flu was that it might have a mutation in its hemagglutinin gene that would allow the precursor protein to be cleaved by enzymes in cells outside the lungs. If so, the flu could invade other body tissues and organs, which might be why it was so deadly. It might be able to infect cells in the brain, for example, causing encephalitis lethargica.

Taubenberger barely allowed himself to hope, as he and Ann Reid carefully put together the sequence of the hemagglutinin gene from the 1918 flu virus. It would be too easy if the first hypothesis they examined turned out to be correct.

But to their disappointment, the cleavage site of the hemagglutinin protein turned out to be perfectly ordinary. On February 16, 1999, they published a paper in the *Proceedings of the National Academy of Sciences* giving the gene's sequence. If the virus could spread to the brain or to other tissues of the body, a hemagglutinin mutation was not the way it did so.

With that idea ruled out, Taubenberger moved on to another popular hypothesis: that mutations in the neuraminidase gene allowed the virus to spread outside the lungs. This idea came from experiments with mice, animals that normally are resistant to influenza viruses. When scientists systematically and repeatedly injected a human flu virus directly into the rodents' brains, eventually the virus's neuraminidase gene mutated and caused a lethal encephalitis. The conclusion was that the 1918 flu could have had a similar mutation, allowing it to grow in the human brain. It was a provocative way of tying together von Economo's disease and the deadliness of the 1918 flu.

The mutation created an indirect way for enzymes in brain cells to split the virus's hemagglutinin protein, accomplishing the same task that a mutation in the hemagglutinin gene might have accomplished. Yet those neuraminidase gene mutations in the viruses infecting mice were unusual and had never been seen in a naturally occurring flu virus. It remained possible that they had occurred in the 1918 flu and that they had been the reason that flu was so deadly, which led Taubenberger and Reid to look at that gene as soon as they finished with the hemagglutinin.

Taubenberger and Reid, however, found no evidence of such a neuraminidase gene mutation. "We have no molecular evidence to support the idea that the virus could get out of the lungs," Taubenberger said. "We've kind of run out of luck looking for the known mutations and so now we're looking for things that are unique" and that could, in the end, have made the virus deadly.

The next step was to examine a hypothesis advanced by Palese.

Palese came upon his idea almost by accident.

He and his colleagues were experimenting with a type of artificial flu viruses—ones they constructed by creating mutations in specific flu genes in the laboratory and then making viruses that had one altered gene while the rest were left intact. They were not trying to make monster flu strains, but instead were working on the technique because it could be important for making vaccines. Scientists could create viruses that would not cause infections—because the deliberately created genetic alterations had defused them—but would elicit immune responses.

As part of this work, Palese and his associates, Dr. Adolfo García-Sastre of Mount Sinai and Dr. Thomas Muster of the University of Vienna Medical School, decided to study a flu virus that was missing a gene, called NS1, that directs the virus to make a protein that lies buried inside the virus particles. No one knew what the NS1 protein did—this would be a way of finding out.

To the scientists' immense surprise, the flu viruses without functioning NS1 genes were able to grow—and kill—a certain strain of mice. These mice lacked the ability to make interferon, a protein made by white blood cells as part of the body's natural defense against viruses. Normally, when a cell is infected with a virus, interferon seeps into the cell and slows the virus's growth, helping to contain the viral infection.

It looked like the NS1 protein of influenza viruses was the virus's way of blocking interferon. If interferon was the body's antivirus missile, interferon was the virus's antiballistic missile.

Palese drew what, to him, was the obvious conclusion: a flu virus with a sort of super-NS1 protein might be extraordinarily deadly because it would prevent interferon from doing its job. Such a flu virus could be a killer. And that, Palese suggested, might be the secret of the 1918 flu.

"I called Taubenberger and said I would like to have the NS1 sequence of the 1918 flu," Palese said.

And if that hypothesis proves wrong?

"I don't want to consider it at this point, but one would have to go back to the drawing board," Palese said.

Taubenberger, however, has his doubt that NS1 will be the final answer. Yes, he is working as fast as he can to find the gene sequence and give it to Palese. But he sees a flaw in the argument.

If the 1918 virus had one simple genetic change that allowed it to evade the body's immunological defense system, why was that change lost in subsequent viruses? Darwinian theory would predict that all future viruses would retain such a mutation since it would confer a huge competitive advantage on the virus.

"If the 1918 NS1 really had the ability to block the interferon response, why would descendant viruses mutate away from such a 'positive' change?" Taubenberger asked.

Taubenberger has his own hypotheses, ones that evade the search for a simple murder weapon that might have armed the 1918 flu.

His first, and preferred, explanation is that the virus was novel, like nothing young people had encountered before, so they had no antibodies to protect themselves. And this virus in particular grew extremely well in human cells, replicating rapidly, so that there were soon enormous numbers of the virus in people's lungs. That would lead to pneumonia and, if large areas of lung cells were killed, to an influx of fluids as well as hemorrhaging. In short, it would lead to the very signs and symptoms of the deadly 1918 flu.

If that is the explanation, Taubenberger said, it is very unlikely that there is a single mutation in the virus that caused the flu to turn into a killer. Instead, he said, there are likely to be "multiple subtle changes in the virus so that all of its gene products interacted well." But, he added, "the problem with that is that since we don't understand the vast majority of these interactions, subtle changes in sequence are not going to be picked up immediately, especially when thinking about only one gene at a time."

On the other hand, Taubenberger said, a flu that is so freakishly perfect for killing people is far out on the edge of what is possible for influenza viruses, meaning that any mutations will make it less deadly. It is a virus in "perfect balance," Taubenberger said, and it is a balance that will tip toward the more mundane type of flu with the tiniest nudge.

When you couple the virus's unprecedented virulence with the fact that anyone who survived it will then be immune to it—meaning that the virus will be forced to mutate or die out—it is not so surprising that the 1918 flu virus seems to have disappeared from the world, Taubenberger observed.

But there is another explanation, one that he thinks is less likely but that he cannot dismiss out of hand. It is that the population that was living in 1918 had a peculiar immunological response to the 1918 flu, one that was prompted by a previous exposure to a different influenza virus, most likely the one that struck in 1890, twenty-eight years before the 1918 flu.

What if babies and children who were exposed to the 1890 flu responded by making abundant quantities of antibodies? Taubenberger asked. And what if the 1918 flu virus had a similar protein on its surface so that the antibodies to the 1890 flu also vigorously attacked the 1918 flu virus? If that happened, the immune system itself, instead of the flu virus, could be causing the flu deaths. In a gross overresponse to the 1918 virus, armies of white blood cells and fluids could rush to the lungs of flu patients. The healthier people were, and the better their immune systems, the more likely they would be to die when they were infected with the 1918 virus.

If that hypothesis is correct, Taubenberger said, it would mean that the 1918 flu virus was not deadly by itself, but instead was deadly because it came at the wrong time. But, he added, "again, the only way to get at this is to find an 1890 virus." And his only hope of that is to search the Army warehouse.

In a sense, it is the ultimate frustration. Scientists have captured the mass murderer, the 1918 flu virus. But they still do not know its murder weapon.

"We definitely have the right suspect, but we do not yet know how the murder was committed," Taubenberger said.

If this story was fiction, the clues would yield a suspect and the suspect would reveal the weapon. But it is science, and science is not always neat and clean. In science, each new finding can open the door to a flurry of new questions.

On the other hand, it may not matter as much if the weapon is found. Medicine has armed doctors with tools that were not available in 1918 to fight a killer influenza strain. Now there are antibiotics that can thwart pneumonia-causing bacteria that swarmed into the lungs of flu victims who were too ill to fight back. No longer will hordes of young people die of bacterial infections that would come in the wake of an influenza virus. And there are now drugs that can temper some influenza infections, possibly softening the blows of a killer flu. With the hemagglutinin gene of the 1918 flu in hand, companies can even make a vaccine that could protect people from that virus if it comes again.

But it is hard to be complacent.

Would a new killer flu look just like the 1918 flu? Or was that virus more of an example of what could happen if a flu virus was perfectly made to be a deadly foe? Will the next terrible influenza virus be a new strain that is, in its own way, ideally made to kill?

Jeffery Taubenberger, for one, thinks we cannot predict what the next lethal flu virus will look like. The only hope we have is vigilant surveillance, keeping a careful eye out for the rough beast whose hour has come at last.

Perhaps, in some innocent encounter in China between a child and a bird, a new killer flu is on its way. Or perhaps, even now, a young man or a young woman has become infected with two different strains of flu viruses. They are mixing together in the person's lungs, their genes reassorting. Emerging from that witches' brew is a new virus, a chimera, that, like the 1918 flu virus, is perfectly suited for destruction.

Perhaps, as we grow almost smug about influenza, that most quotidian of infections, a new plague is now gathering deadly force. Except this time we stand armed with a better understanding of the past to better survive the next pandemic.

# ACKNOWLEDGMENTS

In uncovering the story of the 1918 flu and the search for the killer virus, I especially appreciate the extraordinary assistance I received from Jeffery Taubenberger and Johan Hultin—the long hours they spent discussing their stories with me, their help with my follow-up questions and requests, and their generous provision of documents that were invaluable to me in writing this book. I also want to thank Edwin Kilbourne, John Oxford, and Robert Channock for providing me with articles and letters and for their assistance in my efforts to tell a story that was not just accurate but true. In addition, I had the generous cooperation of scores of scientists who agreed to repeated interviews and provided me with documents and papers and other materials that added facts, perspective, and color to the tale.

I also thank my husband, Bill Kolata, for patiently reading my manuscript and its revisions and for finding documents and data that helped bring the story of the 1918 flu to life.

# NOTES

## 1. THE PLAGUE YEAR

3 **camouflaged German ship** Author's interview with Alfred Crosby, August 28, 1998.

4 **Doane's remarks** *Philadelphia Inquirer*, September 21, 1918.

5 **W-shaped death curves** Gerald F. Pyle, *The Diffusion of Influenza: Patterns and Paradigms* (Totowa, N.J.: Rowman & Littlefield, 1986), p. 50.

6 **origin of word "influenza"** Edwin D. Kilbourne, *Influenza* (New York: Plenum Medical Book Co., 1987), p. 26.

6 **the disease might be cholera . . .** Internet site raven.cc.ukans.edu/~kansite/ww_one/medical/pasrons.htm.

6 **"influenza" in quotation marks** Richard E. Shope, "Old, Intermediate, and Contemporary Contributions to Our Knowledge of Pandemic Influenza," *Medicine* 23 (1944): 422–23.

6 **25 percent of U.S. population** Pyle, *The Diffusion of Influenza*, p. 30.

6–7 **Navy and Army estimates** Alfred W. Crosby, *America's Forgotten Pandemic* (Cambridge: Cambridge University Press, 1989), p. 203.

9 **flu in San Sebastián and Madrid** Richard Collier, *The Plague of the Spanish Lady* (London: Allison & Busby, 1996) pp. 7–8.

10 **Sergeant Acker's letter** Edward M. Coffman, *The War to End All Wars: The American Military Experience in World War I* (New York: Oxford University Press, 1968), p. 80.

10 **why the flu was called Spanish** Shope, "Old, Intermediate, and Contemporary Contributions," p. 419.

10 **Ford and San Quentin** Crosby, *America's Forgotten Pandemic*, p. 18.

11 **the flu in England** Collier, *Plague of the Spanish Lady*, p. 8.

11 **the epidemic in Asia** Richard E. Shope, "The R. E. Dyer Lecture. Influenza. History, Epidemiology, and Speculation," *Public Health Reports*, 73, no. 2 (1958): 168–69.

11 **Grand Fleet** Collier, *Plague of the Spanish Lady*, p. 8.

11 **"It was a grievous business"** Crosby, *America's Forgotten Pandemic*, p. 27.

12 **"had grimly cut its swath"** Pyle, *The Diffusion of Influenza*, p. 41.

12 **forms of the flu** Shope, "R. E. Dyer Lecture," p. 169.

12–13 **the toll in Boston** Crosby, *America's Forgotten Pandemic*, pp. 30, 40.

13 **Roy's letter** *British Medical Journal*, December 22–29, 1979, pp. 1632–33.

14 **William Henry Welch** Crosby, *America's Forgotten Pandemic*, p. 3.

15 **Welch at Fort Devens** Simon Flexner and James Thomas Flexner, *William Henry Welch and the Heroic Age of American Medicine* (Baltimore: Johns Hopkins University Press, 1941), p. 376.

15 **"You will proceed immediately"** Victor C. Vaughan, *A Doctor's Memories* (Indianapolis: Bobbs-Merrill, 1926), pp. 431–32.

15–16 **Deaths at Fort Devens** Crosby, *America's Forgotten Pandemic*, p. 7.

16 **Vaughan's previous experience** Ibid., p. 7.

16 **"hundreds of stalwart young men"** Vaughan, *A Doctor's Memories*, pp. 383–84.

17 **"was quite excited"** Flexner and Flexner, *William Henry Welch*, pp. 376–77.

18 **"our doctors and nurses"** Crosby, *America's Forgotten Pandemic*, p. 48.

18 **draft call canceled** Ibid., pp. 48–49.

18 **flu epidemic in Philadelphia** Ibid., pp. 71–77, and Pyle, *The Diffusion of Influenza*, p. 49.

20 **759 deaths** Bradford Luckingham, *Epidemic in the Southwest, 1918–1919* (El Paso: Texas Western Press, University of Texas at El Paso, 1984), p. 2.

21 **Worldwide spread of flu** Shope, "R. E. Dyer Lecture," p. 169.

21 **"street cars rattled down Bank Street"** Kilbourne, *Influenza*, p. 15.

21 **shortage of coffins in Cape Town** Ibid.

21 **"All the theaters"** Katherine Anne Porter, *Pale Horse, Pale Rider* (New York: Harcourt, Brace & World, 1936), p. 233. The informa-

tion that the novella is autobiographical is from an interview on August 28, 1998, with Alfred Crosby, who communicated with Porter before her death.

21 **"It happened so suddenly"** Kilbourne, *Influenza*, p. 15.

22 **John McCrae** Web site http://www.emory.edu/ENGLISH/LostPoets/McCrae.htn.

22 **"I saw one patient die"** Pyle, *The Diffusion of Influenza*, p. 51.

22 **Camp Sherman victims** Coffman, *The War to End All Wars*, pp. 82–83.

22 **Soldiers chewed tobacco** Ibid.

23 **"no person shall appear"** Luckingham, *Epidemic in the Southwest*, p. 34.

23 **"the ghost of fear"** Ibid., p. 20.

23 **spread of anecdotes** Web site http://www-leland.stanford.edu/~uda/flu.html.

25 **"I am so glad I can help"** Luckingham, *Epidemic in the Southwest*, p. 10.

25 **death of Thomas Wolfe's brother** Thomas Wolfe, *Look Homeward, Angel* (New York: Charles Scribner's Sons, 1929), pp. 452–65.

27 **description of palliative care** Crosby, *America's Forgotten Pandemic*, p. 7.

29 **experiences of Dr. Park** Interview with his daughter, Mrs. William Meade Wheless, April 27, 1998.

30 **Description of Camp Upton** *Order of Battle of the United States Land Forces in The World War. Zone of the Interior: Territorial Departments. Tactical Divisions Organized in 1918. Posts, Camps, and Stations*, vol. 3, part 2 (Washington, D.C.: Center of Military History, United States Army, 1988), p. 796, and *The Medical Department of the United States Army in the World War*, vol. 5: *Military Hospitals in the United States*, prepared under the direction of Maj. Gen. M. W. Ireland, M.D., Surgeon General of the Army, by Lieut. Col. Frank W. Weed, M.C., U.S. Army.

## 2. A HISTORY OF DISEASE AND DEATH

35 **Descriptions of the plague of Athens** Robert Maynard Hutchins, editor-in-chief, *Great Books of the Western World*, vol. 6: *Thucydides: The History of the Peloponnesian War*, translated by Richard Crawley, revised by R. Feetham (Chicago: Encyclopaedia Britannica, 1952), pp. 387–405.

38 **tuberculosis stalked city dwellers** Roy Porter, *The Greatest Benefit to Mankind: A Medical History of Humanity* (New York: W. W. Norton, 1997), pp. 401–42.

38 **the sickness began in China** McNeill, *Plagues and Peoples* (New York: Anchor Books, 1989), p. 175.

39 **Europe's population had tripled** Robert S. Gottfried, *The Black Death: Natural and Unnatural Human Disaster in Medieval Europe* (The Free Press, 1985), p. 15.

39 **Agnolo di Tura** Ibid., p. 45.

40 **Boccaccio** Giovanni Boccaccio, *The Decameron*, translated and with an introduction by G. H. McWilliam (London: Penguin Books, 1972), pp. 37, 52, 53, 56, 197.

43–44 **descriptions of the cholera epidemic in England** R. J. Morris, *Cholera 1832: The Social Response to an Epidemic* (London: Croom Held, 1976), pp. 11, 15, 16, 21–22, 122–23, 145, 197.

43–44 **death can follow two to three hours later** Kenneth Todar, University of Wisconsin, department of bacteriology, on the Web site http://www.bact.wisc.edu/Bact330/lecturecholer.

46 **how Robert Koch discovered the bacteria that cause cholera** Roy Porter, *The Greatest Benefit to Mankind. A Medical History of Humanity* (New York: W. W. Norton, 1997) p. 436, and interview with Gerald Geison, a professor of history at Princeton University.

48 **the only epidemic disease** McNeill, *Plagues and Peoples*, p. 289.

49 **"the gruesome pictures"** Victor C. Vaughan, *A Doctor's Memoirs* (Indianapolis: Bobbs-Merrill, 1926), p. 384.

49 **"I am not going into the history"** Ibid., p. 432.

50 **eminent physicians who went to France** Alfred W. Crosby, *America's Forgotten Pandemic* (Cambridge: Cambridge University Press, 1989), p. 320.

50 **flu in the 88th Division** Ibid., pp. 154–55.

50 **General Pershing's wires** Ibid., p. 157.

50 **General Ludendorff's notion about the influenza epidemic** Robert B. Asprey, *The German High Command at War: Hindenburg and Ludendorff Conduct World War I* (New York: William Morrow, 1991), p. 466.

51 **Dr. Welch's biographer's scant description** Simon Flexner and James Thomas Flexner, *William Henry Welch and the Heroic Age of American Medicine* (Baltimore: Johns Hopkins University Press, 1941), p. 377.

51 **Donald Smythe's two-sentence paragraph** Donald Smyth, *Pershing, General of the Armies* (Bloomington: Indiana University Press, 1986), p. 207.
52 **Crosby's analysis of flu citations** Crosby, *America's Forgotten Pandemic*, p. 314.
52 **Fort Meade memorial service** Ibid., p. 321.
52 **analysis of college textbooks** Ibid., p. 315.
53 **prohibition of public gatherings** Bradford Luckingham, *Epidemic in the Southwest: 1918–1919* (El Paso: Texas Western Press, the University of Texas at El Paso, 1984), p. 2.
53 **Oliver Wendell Holmes quote** Ibid., p. 4.
53 **"The living come in one door"** Ibid.
53 **"The important and almost incomprehensible"** Crosby, *America's Forgotten Pandemic*, p. 311.

## 3. FROM SAILORS TO SWINE

57–59 **Experiments with sailors in Boston and San Francisco** Richard E. Shope, "The R. E. Dyer Lecture. Influenza: History, Epidemiology, and Speculation," *Public Health Reports*, 73, no. 2 (February 1958): 170, 171, and Alfred W. Crosby, *America's Forgotten Pandemic* (Cambridge: Cambridge University Press, 1989), pp. 267, 268, 280, 281, 282.
60 **the three Japanese doctors** T. Yamanouchi, K. Skakami, S. Iwashima, "The Infecting Agent in Influenza," *Lancet*, 196 (June 7, 1919): 971.
61 **"The outstanding fact"** Gerald F. Pyle, *The Diffusion of Influenza: Patterns and Paradigms* (Totowa, N.J.: Rowman & Littlefield, 1986), p. 43.
62 **spread of the disease in Army camps** Richard E. Shope, "Old, Intermediate, and Contemporary Contributions to Our Knowledge of Pandemic Influenza," *Medicine*, 23 (1944): 420.
62 **"In many respects, the epidemiologist"** Ibid., p. 421.
63 **Robert Johnson puzzled over** Shope, "Dyer Lecture," p. 166.
63 **"Towards the end of May"** Ibid., p. 167.
63 **"The present received opinion"** Shope, "Old, Intermediate, and Contemporary Contributions," p. 416.
64 **"What I wish to point out now"** Ibid., p. 417.
64 **Pfeiffer convinced most of the world** Shope, "Dyer Lecture," p. 175.

65 **"About all that can be said"** Ibid., p. 171.

65 **"a vivid sense of humor"** Christopher Andrewes, "Richard E. Shope," *National Academy of Sciences Memoirs*, vol. 60, p. 363.

67 **seeding of swine flu in the Midwest** Crosby, *America's Forgotten Pandemic*, p. 297.

67 **Koen quotes** Shope, "Dyer Lecture," p. 172.

67 **Shope quotes** Shope, "Old, Intermediate, and Contemporary Contributions," pp. 431–33.

72 **Shope's experiment giving the pigs both the filtrate and the bacteria** Ibid.

73 **ferret experiments** Crosby, *America's Forgotten Pandemic*, pp. 286–89.

75 **bacteria were not causing influenza** Shope, "Old, Intermediate, and Contemporary Contributions," p. 434.

75 **could block the flu viruses** Shope, "Dyer Lecture," p. 234.

79 **"In these, then, the added assumption"** Shope, "Old, Intermediate, and Contemporary Contributions," p. 438.

80 **1918 flu and swine flu antibodies** Crosby, *America's Forgotten Pandemic*, p. 304.

80 **statements by Navy Surgeon General and British Grand Fleet** Shope, "Dyer Lecture," pp. 174–75.

## 4. A SWEDISH ADVENTURER

This chapter is based on a series of extended conversations the author had with Johan Hultin in 1998 and 1999, at his home, on the telephone, and through E-mail. She also relied on interviews with others who knew him, such as Jeffery Taubenberger of the Armed Forces Institute of Pathology and John Oxford of London Hospital Medical College and on unpublished diaries and other material, such as newspaper clippings and photographs, that Hultin provided to document his story. In interviews with the author in 1997, Dr. Maurice Hilleman described the military expedition to Alaska and Dr. Nancy Cox of the Centers for Disease Control and Prevention and Dr. Joshua Lederberg of Rockefeller University provided additional personal and technical details.

The history of discoveries about influenza from the 1930s until about 1950 is derived from interviews with Dr. Edwin Kilbourne, Dr. John Oxford, and Dr. Robert Channock and from a timeline in "Influenza: The Virus and the Disease" by Charles H. Stuart-Harris, Geoffrey C. Schild, and John S. Oxford, 1983, p. 264.

## 5. SWINE FLU

121 **death of Private Lewis** Arthur M. Silverstein, *Pure Politics & Impure Science: The Swine Flu Affair* (Baltimore: Johns Hopkins University Press, 1981), p. 4.

122 **Colonel Bartley and Dr. Goldfield's wager** Various authors agree on this bet. See, for example, Edwin D. Kilbourne, *Influenza* (New York: Plenum Medical Book Co., 1987), p. 326.

129 **emergency meeting at the CDC** Silverstein, *Pure Politics*, pp. 24–25. Edwin Kilbourne, in interviews in 1998 and 1999, confirmed this information.

132 Kilbourne described his initial work with the swine flu virus in interviews with the author in 1998 and 1999. He gave a similar account in an interview at the time in Harold Schmeck, "Race for a Swine Flu Vaccine Begins in a Manhattan Lab," *The New York Times*, May 21, 1976, p. C1.

133 **government officials' public announcement** Richard E. Neustadt and Harvey V. Fineberg, *The Epidemic That Never Was: Policymaking and the Swine Flu Scare* (New York: Vintage Books, 1983), p. 20.

134 **"for reasons that are not entirely clear"** Silverstein, *Pure Politics*, pp. 28–29.

135 **"A radically new strain had appeared"** "Vaccine Decision: How the Experts Settled Their Doubts," *Medical Tribune*, April 21, 1976, p. 12.

136 **"All of us would have liked more evidence"** Ibid., p. 1.

136 **"One death, thirteen sick men"** Neustadt and Fineberg, *The Epidemic That Never Was*, p. 22.

137 **20 million doses** Ibid.

137 **"It was clear that we could not say"** Ibid., p. 23.

138 **"I found it difficult to convey"** Kilbourne, *Influenza*, p. 328.

138–139 **"Better to store the vaccine in people"** Neustadt and Fineberg, *The Epidemic That Never Was*, p. 29.

139 **"There was nothing in this for the CDC"** Ibid., pp. 24–25.

140 **"Each was prepared to bet"** Ibid., p. 25.

141 **"the individuals who made the key decisions"** Richard E. Neustadt and Ernest R. May, *Thinking in Time: The Uses of History for Decision Makers* (New York: The Free Press, 1986), p. xii.

141 **"Once differing odds have been quoted"** Ibid., p. 152.

142 **"What 'Alexander's question' forces"** Ibid., pp. 152–53.

142 **Alexander's unimpassioned demeanor** Neustadt and Fineberg, *The Epidemic That Never Was*, p. 27.

143 **"My view is that you should be conservative"** Ibid.
143 **"Suppose there is a pandemic"** Ibid., p. 28.
143 **"This was an opportunity"** Ibid., p. 26.
144 **"our reservations, though voiced"** Letter to the author dated February 12, 1999.
144 **"Stallones summed it up best"** Neustadt and Fineberg, *The Epidemic That Never Was*, pp. 28–29.
145 **"We have not undertaken a health program"** Ibid., pp. 205–6.
145 **"reads as though it were deliberately"** Ibid., p. 31.
145 **Sencer's meeting with Mathews** Ibid., p. 33.
145 **"I presented the issue to Mathews"** Ibid., pp. 34–35.
146 **"Sencer's action memorandum three days later"** Ibid., p. 43.
147 **"the roosters of America"** Ibid., p. 41.
147 **"I think you ought to gamble"** Ibid., p. 42.
149 **"unanimity meant less than Ford assumed"** Ibid., p. 220.
149 **"if you've got unanimity" . . . "I have been advised"** Ibid., p. 46.

## 6. A LITIGATION NIGHTMARE

152 **"ranged across the list"** Richard E. Neustadt and Harvey V. Fineberg, *The Epidemic That Never Was: Policymaking and the Swine Flu Scare* (New York: Vintage Books, 1983), p. 47.
153 **"Some experts seriously question"** Ibid.
154 **"When one of them told me"** *American Journal of Epidemiology*, 110, no. 4 (1979): 523.
154 **Cooper's testimony** Neustadt and Fineberg, *The Epidemic That Never Was*, pp. 49–50.
155 **A/Victoria vaccine to be mixed with swine flu vaccine** Arthur M. Silverstein, *Pure Politics & Impure Science: The Swine Flu Affair* (Baltimore: Johns Hopkins University Press, 1981), pp. 78–79.
155 **Pig farmers complained** Ibid., p. 78.
155 **"There are as many dangers"** Neustadt and Fineberg, *The Epidemic That Never Was*, p. 1.
156 **HEW's analysis of newspapers** Silverstein, *Pure Politics*, p. 84.
156 **Criticism by Sabin and Morris** Ibid., p. 85.
156–157 **Parke-Davis mix-up** Ibid., p. 79.
157 **"It has always seemed to me"** Letter to author, February 2, 1999.
158 **"it is indeed highly questionable"** Silverstein, *Pure Politics*, p. 80.
158 **vaccine field tests** Ibid., p. 83.

158 **Federal officials were convinced** Ibid., p. 90.

159 **"Their worry was"** Neustadt and Fineberg, *The Epidemic That Never Was*, p. 78.

160 **"Never mind that the company"** Ibid., p. 71.

161 **Dr. Neumann's letter** *The New York Times*, September 15, 1976.

162 **"Behind these arguments for indemnification"** Neustadt and Fineberg, *The Epidemic That Never Was*, pp. 72–73.

162 **"We would open every meeting"** Ibid., p. 75.

163 **Merrell would stop production** Silverstein, *Pure Politics*, p. 95.

163 **vaccine not put in vials** Neustadt and Fineberg, *The Epidemic That Never Was*, pp. 84–85.

164 **"This is pioneering"** *Congressional Record*, August 10, 1976, p. 26632.

164 **"asked the drug companies to produce"** Ibid., p. 26796.

164 **Gallup poll** Neustadt and Fineberg, *The Epidemic That Never Was*, p. 91.

165 ***Pittsburgh Press* report** Silverstein, *Pure Politics*, p. 110.

165 **"definitely a possibility"** Neustadt and Fineberg, *The Epidemic That Never Was*, p. 91.

166 **"We are setting up a program"** Ibid., p. 92.

166 **"We know that substances"** Ibid.

166–167 **Dr. Couch's anecdote** Interview with author, May 4, 1999.

167 **40 million had shots** *In re Swine Flu Immunization Products Liability Litigation. Verlin G. Uthank, Plaintiff*, v. *United States of America, Defendant*, Civ. A. No. 78-F-452, United States District Court, District of Utah, 533 F. Supp., 703; 1982, U.S. Dist.

168 **"We felt we were sitting on a time bomb"** Neustadt and Fineberg, *The Epidemic That Never Was*, p. 96.

168 **cases in Alabama and New Jersey** Philip Boffey, "Guillain-Barré: Rare Disease Paralyzes Swine Flu Campaign," *Science*, 194 (January 14, 1977): 155.

168 **CDC search of published papers** Silverstein, *Pure Politics*, p. 118.

169 **29 cases of Guillain-Barré syndrome** Ronald P. Lesser et al., "Epidemiologic Features of Guillain-Barré Syndrome: Experience in Olmsted County, Minnesota, 1935 through 1968," *Neurology*, 23 (December 1973): 1269–72.

169–171 **Kurland on epidemiological problems** Interviews with author, December 1998, January 1999.

171–172 **Dr. Schonberger's account** Interview with author, June 17, 1999.

173 **"I think until you have"** Boffey, "Guillain-Barré," p. 158.

173 **"Problems with diseases"** Neustadt and Fineberg, *The Epidemic That Never Was*, p. xxv.

176–177 Judge Sherman Finesilver's statements are from interviews with the author in December 1998 and January 1999. For his opinions, see also *In re Swine Flue Immunization Products Liability Litigation. Verlin G. Uthank, Plaintiff*, v. *United States of America, Defendant*, United States District Court, District of Utah, Civ. A. No. 78-F-452 (January 4, 1982): 702–27, and *Verlin G. Uthank, Plaintiff-Appellee* v. *United States of America, Defendant-Appellant*, United States Court of Appeals for the Tenth Circuit, no. 82-2272, 732 F.2d 1517 (May 1, 1984): 1517–22, and *In re Swine Flu Immunization Products Liability Litigation. Joseph Lima, Plaintiff* v. *United States of America, Defendant*, United States District Court of Colorado, Civ. A. No. 80-F-16, 508 F. Supp. 897 (February 24, 1981): 897–905.

177–180 **Alvarez case** *In re (Swine Flu Immunization) Products Liability Litigation. Jennie Alvarez, Plaintiff* v. *United States of America, Defendant*, United States District Court, District of Colorado, Civ. A. No. 78-F-1128, 495 F. Supp. (1980): 1188–1208.

182 **no follow-up of cases** Leonard Kurland et al., "Swine Influenza Vaccine and Guillain-Barré Syndrome: Epidemic or Artifact?" *Archives of Neurology*, 42 (November 1985): 1089–90. See also Leonard T. Kurland, "The Role of Epidemiology in Product Liability Litigation with Special Emphasis on the Swine Flu Affair in the United States," presented at the annual meeting of the Canadian Life Insurance Medical Officers Association, in Winnipeg, May 7–9, 1986.

183 **Robert Couch's remarks** Interview with author, May 4, 1999.

183 **Guillain-Barré syndrome in Michigan and Minnesota** T. J. Sfranek, D. N. Lawrence, L. T. Kurland et al., "Reassessment of the Association Between Guillain-Barré Syndrome and Receipt of Swine Influenza Vaccine in 1976–1977: Results of a Two-State Study: Expert Neurology Group," *American Journal of Epidemiology*, 119 (1984): 880–89.

185 **Keiji Fukuda's remarks** Interview with author, December 18, 1998.

## 7. JOHN DALTON'S EYEBALLS

The quotes and information in Chapter 7 are from a series of interviews and E-mails with Dr. Jeffery Taubenberger, Ann Reid, and scientists who could confirm details of some of their statements, including Dr. Edwin Kilbourne and Dr.

Robert Channock of the National Institutes of Health. The interviews took place throughout 1998 and during the first few months of 1999. The author also reviewed the published papers that are referred to in the text.

## 8. AN INCIDENT IN HONG KONG

This chapter is based on extensive interviews that the author conducted with Dr. Nancy Cox and Dr. Keiji Fukuda in late 1998 and early 1999 and on interviews with Dr. Kennedy Shortridge in early 1999. Additional information was obtained from interviews the author conducted with virologists who had followed the bird flu episode, including Dr. Robert Channock and Dr. John LaMontagne of the National Institutes of Health, Dr. Jeffery Taubenberger of the Armed Forces Institute of Pathology, and Dr. Robert Webster of St. Jude Children's Research Hospital in Memphis.

Information on the killing of the chickens in Hong Kong and the quote from Steven Ip—"We will start destroying all the chickens in Hong Kong Island, Kowloon Island, and the New Territories"—is from Elisabeth Rosenthal, "Chickens Killed in Hong Kong to Combat Flu," *The New York Times*, December 29, 1997, p. A1. The quotes from Lessie Wei—"Most of our staff members had never encountered a live chicken" and "There was a learning process. Some are now experts in killing chickens"—are from Elisabeth Rosenthal, "Hong Kong to Inspect Mainland Farms for Bird-Flu Virus," *The New York Times*, December 31, 1997, p. A3.

## 9. FROM ALASKA TO NORWAY

The quotes and details in Chapter 9 are from a series of interviews with the principals in the story, especially Kirsty Duncan, Dr. Jeffery Taubenberger, and Dr. Johan Hultin. The author also interviewed others who were present at key moments, including Dr. Nancy Cox, Dr. John Oxford, Dr. John LaMontagne, Dr. Robert Webster, Dr. Peter Lewin, Dr. Edwin Kilbourne, and Dr. Peter Jahrling. The interviews took place between 1997 and 1999.

The remarks by Esther Oxford are from Esther Oxford, "Secrets of the Grave," *The Independent, The Sunday Review*, September 27, 1998, pp. 14–19.

## 10. MYSTERIES AND HYPOTHESES

282 **Dr. Oxford's recollections** Interviews with author, 1998 and 1999.

292 **von Economo's description of the disease** Constantin von Economo, "Encephalitis Lethargica," *Wiener Klinische Wochenschrift*,

30 (1917): 581–85. I relied on a translation in Robert H. Wilkins and Irwin A. Brody, "Encephalitis Lethargica," *Archives of Neurology*, 18 (1968): 324–28. Additional information on the disease is in Hans Zinser, "The Present State of Knowledge Regarding Epidemic Encephalitis," *Archives of Pathology*, 12 (1965): 271–300, and in Oliver Sacks, *Awakenings* (New York: E. P. Dutton, 1983), pp. 13–23.

295 **"were it not for the slaughtering"** Kennedy F. Shortridge, "The 1918 'Spanish' Flu: Pearls from Swine?" *Nature Medicine*, 5, no. 4 (April 1999): 384. His other quotes and speculations are from interviews with the author in December 1998 and January 1999. For additional information on his views, see Kennedy Francis Shortridge, "Pandemic Influenza: A Zoonosis?" *Seminars in Respiratory Infections*, 7, no. 1 (March 1992): 11–25; Kennedy F. Shortridge, "Is China an Influenza Epicenter?" *Chinese Medical Journal*, 110, no. 8 (1997): 637–41; and K. F. Shortridge, "The Influenza Conundrum," *Journal of Medical Microbiology*, 46 (1997): 813–15.

299–300 **Palese's speculations** Interview with author, April 21, 1999. See also Deborah A. Buonagurio et al., "Evolution of Human Influenza A Viruses over 50 Years: Rapid, Uniform Rate of Change in NS Gene," *Science* 232 (May 1986): 980–82, and William Luytjes et al., "Amplification, Expression, and Packaging of a Foreign Gene by Influenza Virus," *Cell*, 59 (December 22, 1989): 1107–13.

300–302 **Taubenberger's speculations** Interviews with author, 1998 and 1999. A paper on the hypothesis that the neuraminidase gene might explain the virulence of the 1918 virus and von Economo's disease is in Jeffery K. Taubenberger, "Influenza Virus Hemagglutinin Cleavage into HA1, HA2; No Laughing Matter," *Proceedings of the National Academy of Sciences, USA*, 95 (August 1998): 9713–15.

303 **Palese on NS1 gene** In Adolfo García-Sastre, "Influenza A Virus Lacking the NS1 Gene Replicates in Interferon-Deficient Systems," *Virology*, 252 (1998): 324–30.

# INDEX

## A

# B



# C

# I



# J



# K

## N



## O



## P

# T



# U